新编
土木工程专业英语

（建筑工程方向）

（第三版）

钱永梅　蔡靖娓　鞠东蕾◎主　编
谢新颖　梁　锐　翟　莲◎副主编
雷国光◎主　审

化学工业出版社
·北京·

本书是《新编土木工程专业英语（建筑工程方向）》的第三版。在第二版的基础上，编者结合建筑工程专业英语的经验和体会编写成本书。本书由浅入深地介绍了专业英语的基础知识、翻译和写作方法，并选编了部分英文文献，训练了读者的阅读能力和技巧。为了便于读者检索和学习，本书还在书后整理出了词汇表和专业英语常用词缀、常用数学符号的文字表达、土木工程中常用的度量衡和单位换算，增加了实用性。

本书为高等学校土木工程专业（建筑工程方向）学生学习专业英语而编写，也可作为广大从事建筑工程专业、工程管理专业工作并具备一定英语基础的工程技术人员及自学者的参考资料。

图书在版编目（CIP）数据

新编土木工程专业英语．建筑工程方向/钱永梅，蔡靖娓，鞠东蕾主编．—3版．—北京：化学工业出版社，2020.1

ISBN 978-7-122-35418-1

Ⅰ.①新… Ⅱ.①钱…②蔡…③鞠… Ⅲ.①土木工程-英语 Ⅳ.①TU

中国版本图书馆CIP数据核字（2019）第231613号

责任编辑：董 琳　　装帧设计：关 飞

责任校对：王素芹

出版发行：化学工业出版社（北京市东城区青年湖南街13号 邮政编码100011）

印 刷：北京京华铭诚工贸有限公司

装 订：三河市振勇印装有限公司

787mm×1092mm 1/16 印张13¼ 字数346千字 2019年12月北京第3版第1次印刷

购书咨询：010-64518888　　售后服务：010-64518899

网 址：http://www.cip.com.cn

凡购买本书，如有缺损质量问题，本社销售中心负责调换。

定 价：**48.00元**

第三版前言

《新编土木工程专业英语（建筑工程方向）》是供高等学校土木工程专业学生学习的专业英语教材，为了适应教学及读者阅读使用，同时为了适应建筑行业发展中绿色建筑、装配式建筑和建筑工业化的迅速发展，本次修订重点对第一部分进行了改版，对第二～第四部分内容进行了增加和完善，同时对第二版中的部分内容进行了更正。

本教材总体编排内容基本不变，共分四个部分。第一部分为基础知识，包括三个单元：第一单元，主要介绍专业英语的基本特点；第二单元，专业英语的翻译，阐述专业英语的翻译方法和技巧；第三单元，科技论文的写作，介绍科技英语的基本体例和写作基本知识。第二部分为工程方面英文文献的选编。结合专业英语专业性比较强的特点，本部分集中选编了建筑工程方面的英文文献17篇，涉及基本知识、力学、材料学、建筑组成、结构形式、设计原理、结构性能、建筑施工、计算机辅助设计、工程合同、装配式建筑等建筑工程专业各个方面的内容。第三部分为阅读材料，为了给读者进一步学习有关专业英语知识提供方便，进一步扩大本书的知识覆盖面，本书又选编了17篇阅读材料作为辅助材料。第四部分为词汇表，除了汇总了选编的英文文献中的生词和主要专业词汇外，还汇编了最新建筑工程相关规范中的专业术语词汇，使读者能准确掌握专业词汇的标准英文表达。

本教材英文文献选材针对性较强，选材广泛，难度适中，内容结合土木工程专业知识学习特点。同时，为了便于读者使用，本书附录包括专业英语常用词缀、常用数学符号的文字表达、土木工程常用的度量衡和单位换算等内容。

第三版教材由吉林建筑大学钱永梅、蔡婧娓、鞠东蕾任主编，谢新颖、梁锐、翟莲任副主编，崔健、王若竹、金玉杰、田伟、庞平等参编。全书由雷国光教授主审。

由于编者水平有限，书中难免有疏漏和不足之处，恳请广大读者和同行、专家批评指正。

编　者

2019年7月

CONTENTS

Part Ⅰ The Basic Knowledge

Unit 1 The Basic Characters of English for Special Purpose …… 1
1.1 The Characters of Language 语言特点 …… 1
1.2 The Characters of Grammar 语法特点 …… 2
1.3 The Characters of Words and Expressions 词汇特点 …… 3
1.4 The Characters of Structure 结构特点 …… 5
Unit 2 The Translation of English for Special Purpose …… 8
2.1 Introduction 引言 …… 8
2.2 Contrast Between English and Chinese 英汉语言对比 …… 10
2.3 Selecting and Extending the Meaning of a Word 词义选择及引申 …… 12
2.4 Method of Changing the Syntactical Functions 词性的转换译法 …… 14
2.5 Methods of Adding and Omitting 增译和省译法 …… 16
2.6 Translation of Special Sentence Pattern 特殊句型的翻译 …… 19
2.7 Translation of Long Sentence 长句的翻译 …… 23
2.8 Translation of Subordinate Clause 从句的翻译 …… 25
2.9 Translation about Quantity 有关数量的翻译 …… 27
Unit 3 Writing of Scientific and Technical Papers …… 29
3.1 Stylistic Rules of Papers 论文体例 …… 29
3.2 Title and Sign 标题与署名 …… 30
3.3 Abstract 摘要 …… 31
3.4 Writing and Organizing of the Main Text 正文的组织与写作 …… 34

Part Ⅱ Collection of English Literatures about Engineerings

Unit 1 Careers in Civil Engineering …… 48
Unit 2 Modern Buildings and Structural Materials …… 52
Unit 3 Building Types and Design …… 56
Unit 4 Introduction to Mechanics of Materials …… 59
Unit 5 Loads …… 62
Unit 6 Subsoils and Foundations …… 65
Unit 7 Philosophy of Structural Design …… 68
Unit 8 Safety Concepts …… 72
Unit 9 Design Criteria for Tall Building …… 76
Unit 10 Durability at Concrete Structures …… 80

Unit 11 Prestressed Concrete …… 85
Unit 12 Structure Steel …… 89
Unit 13 Earthquake Prediction and Effect of Earthquake on Structures …… 95
Unit 14 Computer-Aided Drafting and Design …… 99
Unit 15 Construction Engineering …… 103
Unit 16 Civil Engineering Contracts …… 108
Unit 17 Composite Structures and Design Philosophy …… 111

Part Ⅲ The Reading Materials

Unit 1 Components of A Building …… 117
Unit 2 Building Materials …… 122
Unit 3 Special Concrete …… 127
Unit 4 The Procedures of Structural Design …… 129
Unit 5 Reinforced Concrete Columns in a Frame …… 131
Unit 6 Structural Reliability …… 133
Unit 7 Bond of Prestressing Tendons …… 135
Unit 8 Cable Structure …… 138
Unit 9 Yield Line Theory of Slabs …… 142
Unit 10 Concrete Operations …… 145
Unit 11 Future Trends in Construction …… 150
Unit 12 Scheduling and Control of Construction …… 152
Unit 13 Contractors' Management Game …… 154
Unit 14 The Construction Process May Be Automated In the Future …… 156
Unit 15 FIDIC Tendering Procedure …… 159
Unit 16 Introduction of Industrialized Construction …… 163
Unit 17 The Commercial Property Investment of Green Building …… 167

Part Ⅳ Words and Phrases

Unit 1 Words and Phrases of Literatures …… 170
Unit 2 Words and Phrases of Engineering Codes …… 191
2.1 Load 荷载 …… 191
2.2 Subgrade and Foundation 地基基础 …… 191
2.3 Timber Structure 木结构 …… 192
2.4 Concrete Structure 混凝土结构 …… 192
2.5 Steel Structure 钢结构 …… 193
2.6 Masonry Structure 砌体结构 …… 193
2.7 Tall Building 高层建筑 …… 194
2.8 Seismic of Structure 结构抗震 …… 194
2.9 Building Engineering 建筑施工 …… 195
2.10 Ground Treatment 地基处理 …… 196

2.11 Subgrade and Foundation and Construction of Water proof
地基基础及防水施工 …… 197
2.12 Construction of Ground and Roof and Decoration
地面、屋面及装饰装修施工 …… 197

Appendix

Appendix Ⅰ Common Affix in English for Special Purpose
专业英语常用词缀 …… 199
Appendix Ⅱ Expression of Common Mathematical Symbol
常用数学符号的文字表达 …… 202
Appendix Ⅲ Matrixing of Unit and Length，Capacity and Weight in Civil Engineering
土木工程中常用的度量衡和单位换算 …… 205

References

Part I The Basic Knowledge

Unit 1 The Basic Characters of English for Special Purpose

专业英语与普通英语、文学英语相比，有许多独特之处。因为专业英语与专业知识紧密联系，除了包含一些数据（data）、公式（formula）、符号（symbol）、图表（diagram and chart）和程序（procedure）等外，在语言、语法、修辞、词汇、体裁等方面都有其独特之处。下面从语言上、语法上、词汇上和结构上对专业英语的基本特点做一个简要介绍。

1.1 The Characters of Language 语言特点

1.1.1 Accuracy 准确

所谓准确，就是要表达准确，要正确理解和分析英语的语法特点与句型，表达上不使用模棱两可的词。从下面的例子，可看出专业英语的语言特点。

［例 1］ Civil engineering offers a particular challenge because almost every structure or system that is designed and built by civil engineers is unique. One structure rarely duplicates another exactly.

土木工程提出了特殊的挑战，因为由土木工程师设计建造的每个结构或系统几乎都是唯一的。一个结构几乎不能完全复制另一个。

1.1.2 Brevity 简洁

专业英语的内容通常包括理论分析、公式推导和研究的目的、范围、方法、步骤、结论等。在不影响表达的前提下，语言应尽可能简洁，避免不必要的润饰和重复。但并不排除会使用复杂句或长句。

［例 2］ The yield criterion for a material is a mathematical description of the combinations of stresses which would cause yield of the material. In other words it is a relationship between applied stresses and strength.

材料的屈服准则指可能导致材料屈服的应力组合的数学表达式。换句话说，它表示的是应力与强度之间的关系。

1.1.3 Clarity 清晰

清晰主要是强调逻辑严谨、概念清晰、关系分明、句子连贯等。

［例 3］ The materials are the basic elements of any building. Building materials may be classified into three groups, according to the purposes they are used for. Structural materials are those that hold the building up, keep it rigid, form its outer covering of walls and roof, and divide its interior into rooms. In the second group are materials for the equipment inside the building, such as the plumbing, heating, and lighting systems. Finally, there are materials that are used to protect or decorate the structural materials.

材料是任何建筑的基本元素。根据使用目的，建筑材料被分成三组。结构材料用来支撑建筑物保持其坚固，形成墙和屋面的外部围护，以及分隔内部房间。第二组材料是建筑物内部的设备，如垂直运输、加热和照明系统。最后一组材料是用于保护和装饰结构的建筑材料。

1.2 The Characters of Grammar 语法特点

1.2.1 非人称的语气和客观的态度，常使用 It... 结构

专业英语所涉及的内容多描述客观事物、现象和规律。由于这一特点，决定了科技人员在撰写科技文献时采用客观和准确的手法陈述被描述对象的特性和规律、研究方法和研究成果等，而不需要突出人。因此，专业英语常常使用非人称的语气做客观的叙述。

[例 1] It is easier to make changes in design and to correct errors during construction (and at less expense) if welding is used.

若采用焊接，则在施工阶段更容易（以更少的费用）修改设计或改正错误。

例句中采用了 It is... 的结构，对某种事情或事实进行客观的描述，没有加入任何的主观色彩。句中的 It 表示 to make... and to correct... construction。

1.2.2 较多使用被动语态

由于专业英语的客观性，决定了它非人称的表达方式。读者或者都知道动作的执行者是谁，或者不需要说明动作的执行者。因此，在专业英语中，较多地使用被动语态。

[例 2] Before any civil engineering project can be designed, a survey at site must be made.

在设计任何土木工程项目之前，必须先进行现场测量。

1.2.3 大量使用不定式、动名词、现在分词和过去分词

专业英语中大量使用不定式、动名词、分词，多数情况下是为了使句子简洁和精练。

[例 3] The total weight being less, it is possible to build much taller buildings.

由于总重量减轻，就有可能建造更高的楼房。

[例 4] The demands for sophisticated analysis, coupled with some serious limitations on computational capability, led to a host of special techniques for solving a corresponding set of special problems.

因为对精细分析的要求，但又受到计算能力的某些严重限制，由此产生了许多特殊方法以解决相应的一系列特殊问题。

1.2.4 较多地使用祈使语气和公式化表达方式

在理论分析和公式推导中常采用 Assume that..., Suppose that..., Let... 等祈使语气表达方式。

[例 5] Suppose that $P=0$ at $x=y$.

假定当 $x=y$ 时 $P=0$。

1.2.5 条件语句较多

条件语句多用于条件论述、理论分析和公式推导中，最常用的是 If... 条件句。

［例 6］ The huge investment in the infrastructure will be erased quickly if proper maintenance and rehabilitation procedures are enforced and funded.

如果合理的养护和修复计划得以资助并实施，就可迅速取消用于基础建设的巨大投资。

［例 7］ If substituting Eq.（1）into（7），we obtain $F=xyz$.

若把方程式(1) 代入式(7)，则得到 $F=xyz$。

1.2.6 长句较多，但一般比较简洁清晰

［例 8］ It is important also that the designer be aware of the method of construction or erection to be employed since，in certain cases，the loading conditions to which a member is subjected during erection may induce a stress condition which exceeds that due to the service loads of the structure.

设计者了解所采用的施工或安装方法也是很重要的，因为在某些情况下，安装过程中杆件承受的荷载所产生的应力可能会超过使用荷载产生的结构应力。

例句中，that 引出主语从句，since 引出原因状语从句；在 since 从句中，包含两个由关系代词 which 引出的定语从句，分别修饰 the loading conditions 和 a stress condition。

1.2.7 省略句较多

为了简洁，有时省略掉句子中的一些部分，如状语从句中的主语和谓语、定语从句中的关联词 which 或 that，从句中的助动词等，但基本不省略形容词、副词。

［例 9］ If not well managed，the procedure for construction may be more expensive.

如果管理不善，这一施工方法还可能更费钱。

常见的省略句型有：

As already discussed	前已讨论
As described above	如前所述
As explained before	前已解释
As indicated in Fig. 1	如图 1 所示
As previously mentioned	前已述及
Where possible	在可能的情况下
If possible	如果可能的话
If so	倘若如此
When（If）necessary	必要时
When needed	需要时
Where feasible	在实际可行的场合

1.3 The Characters of Words and Expressions 词汇特点

1.3.1 专业（Special）词汇和半专业（Semispecial）词汇

每个专业都有一定数量的专业词汇或术语。例如，对建筑结构工程专业，有 slab（板）、beam（梁）、column（柱）、gable（山墙）、roof（屋面）、bearing wall（承重墙）、cavity brick（空心砖）等；对道路和桥梁工程专业（road and bridge engineering），有 pavement（路面）、roadbed（路基）、abutment（桥台）、pier（桥墩）、deck（桥面）等。

专业文献中的专业词汇一般有三类：第一类是纯专业词汇。它的意义很单纯，只有一种

专业含义，有时候则是根据需要造出来的。如 T-beam（T 形梁）、fire-proof brick（耐火砖）、prestressed concrete（预应力混凝土）等。第二类是半专业词汇。它大多是各个专业通用的，在不同的专业领域却可能有不同的含义。如：foundation（基础、基金、创立）、frame（框架、屋架、机座、体系等）、operation（操作、运行、作业、效果等）、load（荷载、加载、装入、输入等）。第三类是非专业词汇。这类词汇是指在非专业英语中使用不多，但却严格属于非专业英语性质的词汇。如：application（应用、用途、作用、申请等）、implementation（实现、执行、运行）、to yield（产生、得出、发出等）等。

1.3.2 词性（Syntactical Functions）转换

专业英语也较多使用了词性的转换。转换后词意往往与原来的词意相关。常见的词性转换类型有：名词→动词、形容词→动词、动词→名词、形容词→名词等。这里有两种情况，一种是词本身可以在句子中充当另一种词类；另一种是在译文中被转换成另一种词类。例如：standard（*n.* 标准）→standardize（*v.* 标准化）；former（*adj.* 前面的）→the former（*n.* 前者）；wide（*adj.* 宽的）→widen（*v.* 加宽）。

1.3.3 词缀（Affix）和词根（Etyma）

由于历史的原因，英语中的很多文字源于外来语，如希腊语、拉丁语、法语、德语、意大利语和西班牙语等。有些词是日常生活中常用的，例如 economical，immigrate，foreword 等；有的则用于某些专门的领域。例如在土木工程领域，有 hydraulics，infrastructure，reliability，specification 等。据有关专家统计，现代专业科技英语中，有 50%以上的词汇源于希腊语、拉丁语等外来语，而这些外来语词汇构成的一个主要特征就是广泛使用词缀（包括前缀 prefix、后缀 suffix）和词根。因此，如果适当掌握一些词缀和词根，就有助于扩大词汇量。

1.3.4 缩写（Abbreviation）、数学符号（Mathematical Symbol）及其表达式（Expression）

在阅读和撰写专业文献时，常常会遇到一些专有词汇或术语、物理量等单位的缩写，或一些政府机构、学术团体、科技期刊和文献等的简称。例如：

Fig.（Figure）	图
Eq.（Equation）	方程（式）
m/s（meter/second）	米/秒
in.（inch）	英寸
Eng.（Engineering）	工程
i. e.（拉丁语 *id est*）	也就是，即
etc.（拉丁语 *et cetera*）	等等
psi.（pounds per square inch）	磅/英寸
Sym.（Symmetry or Symmetrical）	对称
QC（Quality Control）	质量控制
CAD（Computer Aided Design）	计算机辅助设计
RILEM（International Union of Testing and Research Laboratories for Materials and Structures）	国际材料与结构试验研究所联合会
CIB（International Council for Building Research Studies and Documentation）	国际建筑研究及文献委员会
FIDIC（International Federation of Consulting Engineers）	国际咨询工程师联合会

FIP(International Federation of Prestressing)	国际预应力混凝土委员会
FIB(International Federation for Structural Concrete)	国际结构混凝土联合会
ISO(International Organization for Standardization)	国际标准化组织
ECCS(European Convention of Constructional Steelworks)	欧洲钢结构学会
ASCE(American Society of Civil Engineers)	美国土木工程师学会
ICE(Institute of Civil Engineers)	(英)土木工程师学会
CSCE(Canadian Society for Civil Engineering)	加拿大土木工程学会
ACI(American Concrete Institute)	美国混凝土学会
ASTM(American Society for Testing & Materials)	美国材料与试验学会
NIST(National Institute of Standards and Technology)	(美)国家标准与技术协会
EI(Engineering Index)	(美)工程索引

另外，专业文献中也时常会出现数学符号及其公式和文字表达方式。例如：

[例 1] All primed terms are initially assigned to zero for the experiment.

所有右上角带撇的项在实验开始时均赋零值。

[例 2] Substituting Eq. (5) into (2), dropping higher order terms, and removing the prime notation for simplicity, give the linear variable coefficient system $A=Bx$.

将式(5) 代入式(2)，舍去高阶项，且为简便起见去掉项上的撇号，就得到线形变量系数方程组 $A=Bx$。

1.4 The Characters of Structure 结构特点

上述语言、语法和词汇特点属于专业英语“语域分析”的内容。这些内容形成了专业英语的基础。更进一步，还需要了解专业英语在段落及文章层面上的结构特点，了解隐含在语言运用中的逻辑思维过程。这样才有助于把握文章要点和重点，提高阅读和理解能力。

一般在每一自然段落中，总有一个语句概括出该段落的重点。这个语句或在段落之首，或在段落之尾，较少出现在段落中间。若干个自然段落会形成一个逻辑（或结构）段落，用以从不同角度来解说某一层面的核心内容。全篇则由若干个逻辑段落组成，从不同层面来阐述文章标题所表明的中心思想。

仔细阅读下面一篇短文（其中包括对土木工程的一些重要特性的说明)，分析其结构特点，并结合前面提到的语言、语法和词汇的特点，进一步体会专业英语的特点。

Civil Engineering

① Engineering is the practical application of the findings of theoretical science so that they can be put to work for the benefit of mankind. Engineering is one of the oldest occupations in the history of mankind. Without the skills that are included in the field of engineering, our present-day civilization could never have evolved.

② Civil engineering is a branch of engineering that deals with the design and construction of structures that are intended to be stationary, such as buildings and houses, dams, tunnels, bridges, canals, sanitation systems and the stationary parts of transportation systems-highways, airports, port facilities, and road beds for railroads. Among its subdivisions are structural engineering, dealing with permanent structures; hydraulic engineering, dealing with the flow of water and other fluids; and environmental/sanitary engineering, dealing with water supply, water purification, and sewer systems; as well as urban planning and design. The term civil engineering originally came into use to distinguish it from military engineering. Civil engineering dealt with permanent structures for civilian use, whereas military engineering dealt with temporary structures for military use.

③ Civil engineering offers a particular challenge because almost every structure or system that is designed and built by civil engineers is unique. One structure rarely duplicates another exactly. Even when structures seem to be identical, site requirements or other factors generally result in modification. Large structures like dams, bridges, or tunnels may differ substantially from previous structures.

④ An engineer is a member of the engineering profession. The word engineer is used in two ways in English. One usage refers to the professional engineer who has a university degree and an education in mathematics, science, and one of the engineering specialties. Engineer, however, is also used to refer to a person who operates or maintains an engine or machine. An excellent example is the railroad locomotive engineer, who operates a train. Engineers in this sense are essentially technicians rather than professional engineers.

⑤ Engineers must be willing to undergo a continual process of education and be able to work in other disciplines. They must also adapt themselves to two requirements of all engineering projects. First, the system that engineers produce must be workable not only from a technical but also from an economic point of view. This means that engineers must cooperate with management and government officials who are very cost-conscious. Therefore, engineers must accommodate their ideas to the financial realities of a project. Second, the public in general has become much more aware of the social and environmental consequences of engineering projects and of the hidden or delayed hazards in new products, processes, and many other aspects of civil engineering systems.

⑥ Engineers are required to have solid knowledge of mathematics, physics, and chemistry. Mathematics is very important in all branches of engineering. So it is greatly stressed. A current trend is to require students to take courses in the social sciences and the language arts. The work performed by an engineer affects society in many different and important ways, of which he or she should be aware. An engineer also needs a sufficient command of language to be able to write up his or her findings for scientific publications.

⑦ A civil engineer is a member of the civil engineering profession. They may work in research, design, construction supervision, maintenance, or even in sales or management. Each of these areas involves different duties, different emphases, and different uses of the engineer's knowledge and experience.

...

⑧ Much of the work of civil engineers is carried on outdoors, often in rugged and difficult terrain or under dangerous conditions. Surveying is an outdoor occupation, for example, and dams are often built in wild river valleys or gorges. Bridges, tunnels, and skyscrapers under all kinds of weather conditions. The prospective civil engineer should be aware of the physical demands that will be made on him or her.

分析如下。

这篇文章共分 8 个自然段，介绍 Civil Engineering。第一句就对 Engineering 一词进行了定义，因为段①讨论的是更高一层的 Engineering，它就形成第一个逻辑段。

接着，段②解释什么是 Civil Engineering，其结构的特性（to be stationary），分支（subdivision）情况，Civil Engineering 一词的来源（to distinguish it from military engineering）等。对结构的另外一个重要特性（unique），则在段③加以阐述。这样，段②和段③就形成 Civil Engineering 层面的逻辑段。

段④开始定义 Engineer，说明 Engineer 一词的两种用法；在段⑤中，突出强调专业工程师（professional engineers）所应注意的两方面的问题；段⑥则论述工程师应该掌握的知识和技能。这三段均以 Engineer 为对象，形成第三个逻辑段。

从段⑦开始，就具体到 Civil Engineer；对土木工程实施的一个特点（outdoors），在段⑧加以说明，并由此引出对 Civil Engineer 身体素质要求的评述。段⑦和段⑧组成最后一个逻辑段。

全文的逻辑关系是：围绕土木工程这一主体，内容从粗到细（Engineering→Civil Engineering，Engineer→Civil Engineer），分层展开（Engineering→Engineer，Civil Engineering→Civil Engineer）。

Unit 2 The Translation of English for Special Purpose

2.1 Introduction 引言

所谓翻译，就是把一种语言文字的意义用另一种语言文字准确、完整地表达出来。从这个意义上讲，它是使用不同民族的语言交流思想的工具，也是一个复杂的思维过程，包括观察、记忆、理解、分析、综合、联想、判断、选择等多种思维活动，是用另一种语言文字对原作的思想、氛围、风格进行的一次再创造。

专业英语是英语的一部分，但它又具有独特的形式及专用语言。一般来说，在掌握了一定的英语基础之后，人人都可以动手翻译，但译文未必能满足专业人员的要求。因此，专业英语的翻译就要求翻译者在英语、汉语和专业知识等方面都具有良好的素质和修养。真正地掌握专业英语的翻译，应该主要从以下几个方面着手：

① 掌握适当的专业词汇以及专业符号等；

② 学会分析句子结构（尤其是复杂句）及文章结构，透彻体会原文思想；

③ 学会运用适当的翻译方法和技巧，在忠实于原文的基础上，按照汉语的习惯及专业习惯等将原文准确地表达出来。

2.1.1 Standards of Translation 翻译的标准

翻译的任务在于准确而完整地介绍原文的思想内容，是读者对原文的思想内容有正确的理解。要解决这个问题，就需要有一个共同遵守的翻译标准来衡量译文的质量，来指导翻译的实践。因此翻译标准是衡量译文质量的尺度，又是翻译实践所遵循的原则。

对于翻译的标准，一个比较统一的观点是：信、达（或顺）、雅。“信”是指准确、忠实原作；“达”是通达、顺畅；“雅”是文字优美、高雅。由于专业英语本身注重表现技术问题的科学性、逻辑性、正确性和严密性，所以，专业英语的翻译标准更侧重于：“信”和“达”。

[例 1] The importance of building modern installation can not be overestimated in the economic development.

直译为：在经济发展中，修建现代化设施的重要性不能过分估计。

应译为：在经济发展中，修建现代化设施的重要性无论怎么估计也不过分。

在原文中，over 这种复合词在与 cannot 连用时相当于 cannot... too...，表示“无论如何……也不过分”。直译显然误解了英语的这种特有的表达方式。英语中有许多词存在这样的情况，这是翻译中必须注意的问题。

[例 2] A novel solution to car which runs out of control into bridge abutments and the like had become popular in North American although not yet in Europe.

直译为：对于汽车失去控制撞到墩柱上等类似的问题，尽管在欧洲还未找到解决的办法，然而在北美已经有了新的很普遍的解决办法。

应译为：对于如何避免汽车在失去控制时撞到墩柱上或别的类似的物体上，已经有了一种新的解决办法。这种办法在北美已普遍使用，然而在欧洲却未能做到这一点。

原文的意思是说这种 novel solution 在北美已经 popular 而在欧洲却没有 popular，直译中处理为“欧洲还未找到解决的办法”，显然理解有误。

从以上两个译例可以看到，翻译一定要在准确透彻理解原文的基础上才能进行，切不可不求甚解，想当然而译之。“信”对翻译而言极其重要。然而，“达”是指译文的语言形式应该符合汉语的规范，即翻译时要考虑到汉语的行文习惯和表达方式。译文不顺主要表现在语句的欧化上，逐字死译、生搬硬套。

［例 3］ Grouting of the tendons usually follows the freedom of the ducts from obstruction.

直译为：给钢束灌浆通常跟随在孔道畅通无阻之后。

应译为：钢束灌浆之前，孔道应畅通无阻。

上面译文如果过于拘泥于原文的形式，读起来不仅别扭，而且费解。有时，为了符合汉语行文习惯，需要运用一定的翻译技巧进行适当的变通。

从这些例句可以看到，“信”与“达”是辩证统一的：“信”是“达”的基础。不忠实的译文再通顺也毫无意义；“达”是“信”的保证，不通顺的译文又无疑会影响到译文的质量。因而翻译中必须遵循“信”与“达”相结合的原则。

2.1.2 Process of Translation 翻译的过程

翻译的过程是正确理解原文和创造性地用另一种语言再现原文的过程，大致可分为阅读理解、汉语表达和检查校核等阶段。

(1) 阅读理解

阅读理解阶段是翻译过程的第一步，也是重要的阶段。阅读理解主要是通过联系上下文、结合专业背景进行的。通常应注意两个方面：一是正确地理解原文的词汇含义、句法结构和习惯用法；二是要准确地理解原文的逻辑关系。

［例 4］ Foundation are classified as “rigid” or “flexible”, depending on how they distribute loads.

直译为：基础被分为“坚硬的”或“柔韧的”，这主要取决于它们怎样传递荷载。

应译为：按照传递荷载的情况，基础可分为“刚性的”或“柔性的”。

rigid 译为“坚硬的”，意思上固然不错，但不符合专业术语行文的习惯，应改译为“刚性的”方为妥当。由此可见，在选择词义时，必须从上下文联系中去理解词义，从专业内容上去判断词义。

(2) 汉语表达

表达阶段的任务就是译者根据其对原文的理解，使用汉语的语言形式恰如其分地表达原作的内容。在表达阶段最重要的是表达手段的选择，同一个句子的翻译可能有好几种不同的译法，但在质量上往往会有高低之分。试比较下面的译例。

［例 5］ Action is equal to reaction, but it acts in a contrary direction.

译文一：作用相等于反作用，但它在相反的方向起作用。

译文二：作用与反作用相等，但作用的方向相反。

译文三：作用力和反作用力大小相等，方向相反。

译文一由于拘泥于原文结构，语言不够简练通顺；译文二虽然不错，但不如译文三；译文三完全摆脱了原文形式的束缚，并选用四字结构，使译文准确贴切，简洁有力。

(3) 检查校核

理解和表达都不是一次完成的，往往是逐步深入，最后达到完全理解和准确表达原文的内容。因此，在翻译初稿完成之后，需反复仔细校对原文和译文，尽可能避免漏译、误译。

［例 6］ Theoretically, it may be used for either statically determinate or indeterminate structures, although for practical purposes the method is limited to determinate structures because its use requires that the stress resultants be known throughout the structure.

译文：理论上，这个方法既可用于静定结构，又可用于非静定结构，但在实际应用中，它只限于静定结构，因为用这种方法时，要求知道整个结构的应力合力。

翻译时，既要分析句子的结构，又要考虑逻辑关系，同时要保证没有漏译或误译的现象。由此可见，校核对翻译而言也是非常重要的，尤其在专业英语翻译中，要求高度准确，其中的术语、公式、数字较多，稍有不慎就会造成谬误。

2.2 Contrast Between English and Chinese 英汉语言对比

在翻译中，进行英汉两种语言的对比是十分重要的，特别是比较两者的相异之处。通过对比，能够较为准确地掌握各自不同的特点，这对具体的翻译实践大有帮助。

2.2.1 Contrast of Words and Phrases 词汇的对比

英汉词汇的对比主要是从英语的词义、词的搭配和词序来比较其在汉语中的对应情况，看其对应的程度以及具体使用时会发生怎样的变化。

(1) 词义方面

英语词汇意义在汉语里的对应情况，大致有四种情况。

① 词汇意义一一对应，即对于一些已有通用译名的专有名词和术语等，英汉词汇的意义完全相同。如：civil engineering 土木工程（学）；flexible foundation 柔性基础。

② 英语词汇意义比汉语更广，如：material 比汉语“材料”意义更广；straight 比汉语“笔直”意义更广。

在这种情况下，英语中的词汇与汉语中的词汇在词义上只有部分对应，在意义上概括的范围有广义与狭义之分。例如，material 一词还有物质、剂、用具、内容、素材、资料等词义。翻译时，对这类词要仔细掂量、认真推敲。

③ 英语词汇意义不及汉语广，如：road 不及汉语的“道路”的意义广；car 不及汉语的“汽车”意义广。

这种情况正好与第二种情况相反，例如，中文中“汽车”一词泛指公路车辆，而 car 一词则专指轿车。

④ 英语词汇与相应汉语词汇部分对应，两者的意义都有彼此不能覆盖的部分，如：book 书；state 国家；do 做。

这种情况在英语和汉语词汇关系中是最普遍的，而且也是最难处理的。所以，翻译时特别要注意。

从以上列举的四种情况可以看出，翻译绝不是填充，决不能用某一固定的汉语词去填充某一固定的英语词。一个词的具体意义，只有联系上下文才能确定，如果离开了上下文，孤立地译一个词就很难确切地表达该词的真正含义。

(2) 词的搭配

英语和汉语在词的搭配能力方面往往有差异。如 reduce 基本词义是“减少”，但其搭配能力很广，翻译时需酌情选择适当的汉语词汇。例如：

reduce speed	减低车速
reduce to powder	粉碎
reduce the temperature	降低温度
reduce the time	缩短时间
reduce construction expense	削减工程开支
reduce the scale of construction	缩小工程规模
reduce the numbers of traffic accidents	减少交通事故

英语中一个词可能会有很多意义，翻译时需要注意按汉语的习惯合理选择相应的搭配。

［例 1］ Two or more computers can also be operated together to improve performance or system reliability.

可同时操作两台及以上的计算机，以改善其性能或提高系统的可靠性。

(3) 词序方面

英语和汉语句子中的主语、谓语、宾语和表语的词序大体上是一致的，而定语和状语的位置则有同有异，变化较多。

① 定语的位置 英语中单词作定语时，通常位于所修饰的名词前，但也有少数单词要求后置。汉语的单词作定语一般都前置。如：

movable span	活动跨
journey speed	运行速度
something important	重要的事情

英语中短语作定语一般位于所修饰的名词之后，汉语通常需要前置，间或也有后置的情况，主要看汉语的习惯。如：a building project of tall apartment houses 高层公寓大楼的建筑项目；one of the common defects in concrete maintenance 混凝土维修中普遍存在问题之一。

② 状语的位置 英语中单词作状语，其位置有三种情况：修饰形容词或其他状语时要前置，修饰动词时可前置也可后置，表示程度的状语在修饰状语时通常前置，但也有后置的情况，在汉语中状语一般都需前置。

英语中短语状语可放在被修饰的动词之前或之后，甚至可插入情态动词（或助动词）与实义动词之间。译成汉语时，通常需放在所修饰的动词之前，但也有后置的情况，这要视汉语的习惯而定。

［例 2］ The forces keeping the beam straight must, by a fundamental law of static, equal the load tending to fold it up.

根据静力学原理，使梁保持平直的力必定等于将其压弯的荷载。

［例 3］ To the extent possible, the foundation concrete is placed keeping the excavation dry.

应尽可能在保持基坑干燥的情况下灌筑基础混凝土。

2.2.2 Contrast of Syntax 句法的对比

英汉句法对比主要是指句子结构和句序的比较分析。

(1) 句子结构

英语和汉语在句法结构上有许多不同之处，因而，表达一个相同的意思所运用的表现手法也不尽相同。英语常常使用各种连词、关系代词和关系副词来表达分句以及主句与从句之间的各种关系，而汉语则主要借助词序以及词与短语之间的内在逻辑关系来连接并列复合句和偏正复合句。英译汉时虽然在有些情况下不需要转换句子结构，但很多情况却必须进行这种转换。英汉句子结构转换大致有以下五种情况。

① 英语简单句结构转换成汉语复合句结构。

［例 4］ Considered from this point of view，the question will be of great importance.

如果从这点考虑，这个问题就十分重要。（英语简单句—汉语偏正复合句的假设句）

② 英语复合句结构转换成汉语简单句结构。

［例 5］ Water power stations are always built where there are very great falls.

水力发电站总是建在落差很大的地方。（英语状语从句—汉语简单句）

［例 6］ It is essential that civil engineering students have a good knowledge of mechanics.

学土木工程的学生掌握力学知识是极为重要的。（英语主语从句—汉语简单句）

③ 英语复合句结构转换成汉语不同的复合句结构。

［例 7］ Electronic computer，which have many advantages，can not carry out creative work.

虽然电子计算机有许多优点，但它们不能进行创造性的工作。（英语的主从复合句—汉语转折偏正复合句）

④ 英语的倒装句转换成汉语正常句序。

在英语里，倒装主要是考虑到上下文或语气上的需要，以突出中心。汉语一般不用倒装结构，故英译汉时常常需做适当改变。

［例 8］ Then comes the analysis（or computation）of internal forces.

接下来进行内力分析（或计算）。

⑤ 英语被动结构转换成汉语主动结构，反之亦然。

［例 9］ Soil mechanics and soil stabilization techniques have been used in the construction of footings for buildings.

在建筑的基础施工中，已采用了土力学和土壤稳定技术。

（2）句序

句序是指复合句中主句和从句的顺序。英语和汉语的句序对比，实际上就是比较英语复合句和汉语复合句中按时间和逻辑关系叙述的顺序。

① 时间顺序　英语复合句中，表示时间的从句可以置于主句之前或之后，叙述顺序很灵活，汉语则按照发生的时间顺序叙述，而且汉语的时间通常位于句首。

［例 10］ One must inevitably touch upon the technical aspects when discussing the parking problems，and means of tacking it.

在讨论汽车停放问题和解决办法时，必然会触及与该问题有关的技术环节。

英语复合句中有时包含两个以上的时间从句，各个时间从句的次序也比较灵活，汉语则通常按照事情发生的先后安排其位置。

② 逻辑顺序　英语复合句若是表示因果关系或条件与结果关系，其叙述顺序比较灵活，原因从句或条件从句可以位于主句之前或之后，而汉语中大多是原因或条件在前，结果在后。

［例 11］ This time no one was killed or injured in the accident，for great attention was paid to safety.

由于安全问题受到重视，这次事故中无人身伤亡。

2.3 Selecting and Extending the Meaning of a Word 词义选择及引申

在翻译时，经常会出现一个英语单词对应多个汉语意思，或某些词在词典上找不到适当的词义，难以确切表达原意，甚至造成误解。所以，应根据上下文和逻辑关系，选择恰当的词义，或从其基本含义出发，进一步加以引申。

2.3.1 Selecting the Meaning of a Word 词义的选择

通常，现代英语中一词多类、一词多义的现象特别普遍。同一个词往往属于几个词类，具有不同的意思。因此，在翻译时需要准确选择词义，引申词义，还要注意词类的转译问题。

(1) 与词的语法特征有关

① 词性不同，词义有别。

［例 1］ Using prestress to eliminate cracking means that the entire cross section (rather than the smaller cracked section) is available to resist to bending.

利用预应力来避免裂缝的出现意味着整个截面（而不是开裂后的较小截面）可以抗弯。(prestress 用作名词)

［例 2］ When a curved tendon is used to prestress a beam, additional normal force develops between the tendon and the concrete because of the curvature of the tendon axis.

当采用曲线钢筋束来对梁施加预应力时，由于预应力钢筋束轴线的弯曲影响，在钢筋和混凝土之间会产生附加径向压力。(prestress 用作动词)

② 名词单复数、可数与不可数引起词义改变。

［例 3］ As a result of those economies, many of our most important new projects in other fields became possible.

由于采取了这些节约措施，我们在其他方面的许多最重要的新工程才得以实施。

economy 单数形式既可作“经济”“经济制度”解，又可作“节约”解，但复数形式则是指具体的“节约”措施，不能译为“经济”。

③ 普通名词与抽象名词意义的转变。

［例 4］ Beijing was the first permanent settlement.

北京是最早的永久性居住地。(settlement 用作普通名词)

［例 5］ Enormous stretches of arable land in the central western region are still awaiting settlement.

中西部地区还有大片可耕地有待于垦拓。(settlement 用作抽象名词)

(2) 与词的搭配有关

同一个词、同一类词在不同场合具有不同含义，必须根据上下文的联系及词的搭配关系或专业知识来理解和确定词义。

［例 6］ The works of these watches are all home-produced and wear well.

这些表的机件均系国产，耐磨性好。

［例 7］ Bridges are among the most important, and often the most spectacular, of all civil engineering works.

桥梁是土木工程建筑中最为重要的一种，也往往是最为壮观的一种。

(3) 与汉语表达有关

［例 8］ The gears work smoothly.

齿轮运转灵活。

［例 9］ Statesmen have always worked for peace.

政治家们一直在为和平努力。

2.3.2 Extending the Meaning of a Word 词义的引申

词义引申时，往往可以从词义转译、词义具体化、词义抽象化和词的搭配四个方面来考虑。

（1）词义转译

［例 10］ The choice of material in construction of buildings is basically between steel or concrete，and the main trouble with concrete is that its tensile strength is very small.

钢材和混凝土是建筑施工的基本材料，混凝土的主要缺点是抗拉强度很低。

（2）词义具体化

英译汉时，根据汉语的表达习惯，把原文中某些词义比较笼统的词引申为词义比较具体的词。

［例 11］ There are many things that should be considered in any engineering planning decision.

在任何工程规划的决策中，有许多因素应当考虑。

（3）词义抽象化

英译汉时，有时需要根据汉语的表达习惯把原文中词义比较具体的词引申为词义比较抽象的词。

［例 12］ We have progressed a long way from the early days of aerial surveys.

航空测量自从出现以来，已经有了很大的发展。

（4）词的搭配

［例 13］ In order to get a large amount of water power，we need a large pressure and a large current.

为了大量的水力，我们需要高的水压和强的水流。

2.4 Method of Changing the Syntactical Functions 词性的转换译法

在翻译的过程中，原文中有些词需要转换词性才能使译文通顺自然。词性转译主要有以下四种情况。

2.4.1 Changing into Verb 转译成动词

英语同汉语相比，英语句子中大多数只有一个谓语动词，而汉语动词用得比较多。在该使用汉语动词的场合英语往往会用介词、分词、不定式、动名词或是抽象名词等来表达。

（1）介词动词化

许多含有动作意味的介词，如 across，past，toward 等，译成汉语时通常转译成动词，一些仅表示时间、地点、方式的介词如 in，at，on 等，虽然没有动作意味，但译成汉语时根据汉语的行文习惯有时也需转译成动词。

［例 1］ Mechanical stabilization is still considered of great value in construction of the United States.

在美国施工中，机械稳定法仍然被认为具有很大的价值。

（2）名词动词化

英语中有大量从动词派生的名词和具有动作意味的名词，这类名词在英译汉时常能转译成汉语动词。

［例 2］ This giant entertainment building is under construction.

这座大型娱乐建筑正在兴建。

（3）形容词动词化

英语中表示知觉、感觉、情感、欲望等心理状态的形容词在系动词后作表语时，常常可转译成汉语动词。

［例 3］ Steel is widely used in engineering，for its properties are most suitable for construction purposes.

钢材广泛地用于工程中，因为它的性能非常适合于建筑。

（4）副词动词化

英语中有些副词本身含有动作意味，例如：on，back，off，in，behind，over，out 等。这些副词在英译汉时往往需译成动词。

［例 4］ An exhibition of new building materials is on there.

那里正在举办新型建筑材料的展览会。

2.4.2 Changing into Noun 转译成名词

（1）动词名词化

英语中有很多名词派生的动词和由名词转用的动词，在英译汉时不易找到适当的汉语对应词，因而常将其转译成汉语的名词。

［例 5］ These cracks，however，must be closely watched，for they are constantly being attacked by unfavorable environments.

由于经常受到不利环境因素的侵蚀，这些裂缝必须加以密切观察。

（2）形容词名词化

英语形容词转译成名词大致有三种情况：一是有些形容词加上定冠词表示某一类人或事，汉译时可译成名词，如 the rich（富人），the poor（穷人）等；二是英语的关系形容词在汉语里没有对等词，汉译时常作名词处理，如 ideal structure（理想结构）等；三是科技英语中往往习惯用形容词来表示物质的特性，汉语却习惯用名词，通常在这类形容词后加上“度”“性”等词而转换为汉语名词。

［例 6］ Of those stresses the former is compressive stress and the latter is tensile stress.

在两种应力当中，前者是压应力，后者是拉应力。

（3）副词名词化

英语中由名词派生的副词时常可译成名词，少数不是名词派生的副词有时也可译成名词。

［例 7］ Structural drawings must be dimensionally correct.

结构图的尺寸必须准确。

2.4.3 Changing into Adjective 转译成形容词

（1）名词形容词化

形容词派生的名词，以及带有不定冠词或介词 of 作表语的抽象名词在汉译时可以译成形容词。

［例 8］ The methods of prestressing a structure show considerable variety.

对结构施加预应力的方法是多种多样的。

［例 9］ This experiment is an absolute necessity in determining the best water-cement ratio.

对确定最佳水灰比而言，这次实验是绝对必需的。

(2) 副词形容词化

当英语动词或形容词汉译时名词化后，修饰该动词或形容词的副词也会相应形容词化。

［例 10］ It is a fact that no structural material is perfectly elastic.

事实上，没有一种结构材料是完全的弹性体。

2.4.4 Changing into Adverb 转译成副词

(1) 名词副词化

［例 11］ We find difficulty in solving this problem.

我们觉得难以解决这个问题。

(2) 形容词副词化

英语名词转译成动词时，修饰名词的形容词自然就转译成副词。另外，由于英汉两种语言的表达习惯不同，英语形容词有时需转译成副词。

［例 12］ Engineers have made a careful study of the properties of these structures.

工程师们仔细研究了这些新型结构的特性。

由于句中词性的转化，相应地产生了句子成分的转换，即原文句子中的某一语法成分（主语、谓语、宾语、表语、定语、状语等）改译成另一种语法成分。

［例 13］ Attempts were made to find out measures for reducing construction expenses.

曾试图找到削减工程开支的措施。（主语转换成谓语）

［例 14］ The test results are in good agreement with those obtained by theoretical deduction.

试验结果与理论推导者完全一致。（表语转换成谓语）

2.5 Methods of Adding and Omitting 增译和省译法

翻译时对原文内容不应该做任何删节或增补。但由于两种语言的表达方式不同，把原文信息译成译文信息时，常常需要删减或增添一些词。这样做并不损害原意，反而可以使译文更为通顺，意思更为清楚。这种省略和增补不仅是许可的，而且常常被看成是一种翻译的技巧。

2.5.1 Methods of Adding 增译法

增译法是在翻译时根据句法上、意义上或修辞上的需要增加一些无其形而有其意的词，以便能更加忠实通顺地表达原文的思想内容。当然，增词不是随意的，而是基于汉英两种语言表达方式的差异，增加一些词，以使译文忠信流畅。

(1) 根据句法上的需要

通常，在英语中需要省略的句子成分，在翻译中需要补出，这样才能符合汉语的习惯。

［例 1］ Hence the reason why regulations to control parking in towns are so often viewed with suspicion by Chambers of Commerce.

因此，这正是商会总是以怀疑的眼光对待城市管理停车条例的原因。（句子是省略句，在 hence 后省略了 that is，翻译中需要补出）

(2) 根据意义上的需要

① 增加量词和助词　英语没有（或省略）量词、助词（着、了、过、已经）等，汉译时应该根据上下文的需要增补。

［例 2］ This building was at last finished with the cooperation of all our staffs.

在全体员工的合作之下，这个建筑终于完工了。

② 增加表示复数和时态的词　汉语名词没有复数的概念，动词没有时态的变化，翻译时有必要增加表示复数和时态的词，有时候甚至要添加表示时间对比的词。

［例 3］ Important data have been obtained after a series of experiments.

在一系列的试验之后，得到了许多重要的数据。

［例 4］ The arch structure used to be widely applied to engineering construction. It never has been out of use and never will.

过去，拱结构在建筑施工中被广泛采用。现在它们仍然没有过时，将来也不会。

③ 增加抽象名词　在含有动作意义的抽象名词之后增加"作用""现象""效用""方案""过程""情况""设计""变化"等词，以表示具体概念。

［例 5］ Oxidation will make iron and steel rusty.

氧化作用会使钢铁生锈。

④ 增加动词　根据意义的需要，可以在名词或动名词前后增加动词。通常增加的汉语动词有："进行""出现""产生""引起""发生""遭遇""使"等。

［例 6］ Testing is a complicated problem and long experience is required for its mastery.

进行试验是一个复杂的问题，需要有长期的经验才能掌握。

⑤ 增加解说性词　当英语的某些词单独译出意思不明确时，可在其前增加解说性词使译文意思明确。

［例 7］ Air pressure decreases with altitude.

气压随海拔高度的增加而下降。

⑥ 增加概括性词　在句子中有几个并列成分时，可在其后增添表示数量概念的概括词，达到一定的修辞效果。

［例 8］ A designer must have a good foundation in statics, kinematics, dynamics and strength of materials.

设计人员必须在静力学、运动学、动力学和材料力学这四个方面有良好的基础。

(3) 根据修辞上的需要

英译汉时，有时需要在译文中增加一些起连贯作用的词，主要是连词、副词和代词，以达到使句子连贯、行文流畅的修辞目的。

［例 9］ The Japanese have developed a new type of machine called moles, which can bore through soft and hard rock by mechanical means.

日本人已研制出一种名叫鼹鼠掘进机的新型机械，这种掘进机使用机械方法既可挖掘软岩又可挖掘硬岩。

［例 10］ It is necessary that the calculations should be made accurately.

计算要精确，这一点是很有必要的。

(4) 重复原文中出现过的词

英语中常常会有几个名词共用一个动词，几个形容词共用一个中心词，或是为了避免重复用代词替换先行词等现象。翻译时需要重复原文中重要的或关键的词，以期达到使译文清楚或是强调的作用。另外，有时也将多宾语、状语或表语的动词采用不同的形式分别译出，以便于行文。

［例 11］ An alternative way to use reinforcement is to stretch it by hydraulic jacks before the concrete is poured around it.

另一种方法是先用液压千斤顶把钢筋拉长，然后在钢筋周围浇灌混凝土。（重复代词所指代的对象）

［例 12］ A synthetic material equal to that alloy in strength has been created，which is very useful in civil engineering.

一种在强度上和那种合金相等的合成材料已经制造出来了，这种合金材料在土木工程中很有用。（重复关系代词所指代的先行词）

［例 13］ A body may be exposed to one constant stress，or to variable stress，of even to compound stress，that is where several stresses act on it at the same time.

一个物体可能经受一个不变的应力，或者经受一个变化的应力，甚至可能经受复合的应力，即同时有几个应力作用在它上面。（重复带多个宾语的动词）

［例 14］ Ice is the solid state，water the liquid state，and water vapor the gaseous state.

冰是固态的，水是液态的，而水蒸气是气态的。（重复句中省略的部分）

［例 15］ Also there has been a concreted effort to modernize and increase space，facilities，equipment，and supporting materials used in science teaching.

而且，大家协同努力，从而扩大了场地，发展了机构，增添了设备和科学教育用的辅助材料，并使之现代化。

2.5.2 Methods of Omitting 省译法

所谓省译就是将原文中的某些词语略去不译。在不损害原文内容的基础上，删去一些不必要的词语，会使行文更加简洁明快，充分体现科技文献的一大特点。总的来说，英译汉时省译的现象比增译的现象相对更多。比如说，冠词属于英语里出现频率较高的词，但在汉语里没有，一般可以不译；而有的词如介词、连接词和代词等，在英语里出现频率较高，但汉语则可以通过借助语序表达逻辑关系，所以这几类词有时可以省略。

(1) 省略冠词

［例 16］ The memory is the important part of a computer system.

存储器是计算机系统中的重要组成部分。

当然，在英语一些词组中，冠词的存在使词组的意义发生了很大的变化，所以要另加注意。例如，out of the question（毫无可能，不值得考虑）和 out of question（毫无疑问，不成问题），冠词虽不用特别译出，但是词组的意思刚好相反，这一点在翻译时要体现出来。

(2) 省略代词

［例 17］ If you know the internal forces，you can determine the proportion of members.

如果知道内力，就能确定构件尺寸。

(3) 省略介词

［例 18］ The critical temperature is different for different kinds of steel.

不同类的钢，其临界温度各不相同。

(4) 省略动词

[例 19] All kinds of excavators perform basically similar function but appear in a variety of forms.

各种挖土机的作用基本相同，但形式不同。

(5) 省略连词

[例 20] If there are no heat-treatment, meals can not be made so hard.

没有热处理，金属就不会变得如此坚硬。

[例 21] Up and down motion can be changed to circular motion.

上下运动可以改变为圆周运动。

(6) 省略名词

介词 of 前表示度量意义的名词有时可以省略不译。

[例 22] Different kinds of matter have different properties.

不同物质具有不同的特性。

(7) 省略意义上重复的词

英语中常用 or 引出同位语，这些同位语有的可分别译出，有的具有相同的译名，只能译出一个，省略一个。有时句子里个别词与其他词意义重复，翻译时也应予以省略。

[例 23] The mechanical energy can be changed back into electrical energy by means of a generator or dynamo.

机械能可利用发电机在转变成电能。(省略同位语 dynamo)

2.6 Translation of Special Sentence Pattern 特殊句型的翻译

科技英语中，经常出现被动句型、否定句型、强调句型等。这些句型都有其自身特点，往往和汉语句型有一些不同，在翻译时容易造成错误，因而要特别注意。

2.6.1 Passive Sentence Pattern 被动句型

与汉语相比，英语中被动语态使用的范围要广泛得多。凡是出现以下情况，英语常用被动语态：不必说出行动的行为者；无从说出行动的行为者；不便说出行动的行为者等。英语被动语态的句子译成汉语时，很多情况下都可译成主动句，但也有一些可以保持被动语态。

(1) 译成汉语主动句

① 原文中的主语在译文中仍作主语，将被动语态的谓语译成“加以……”“是……的”等。

[例 1] Distances between elevations are measured in a horizontal plane.

高程之间的(斜)距离是用其水平投影来测量的。

② 原文中的主语在译文中作宾语，将英语句译成汉语的无主语句，或加译“人们”“我们”“大家”“有人”等词作主语。

[例 2] Attempts are also being made to produce concrete with more strength and durability, and with a lighter weight.

目前仍在尝试生产强度更高、耐久性更好，而且质量更轻的混凝土。(无主语)

③ 用英语句中的动作者(通常放在介词 by 后)作汉语句中的主语。

[例 3] The top layers was bound together more firmly by mixing the crushed rock with asphalt.

用沥青掺拌碎石能使表层更坚固地黏结在一起。

④ 将英语句中的一个适当成分译成汉语句中的主语。

［例 4］ Much progress has been made in civil engineering in less than one century.

不到一个世纪，土木工程学取得了许多进展。

(2) 译作汉语被动句

原句中的主语仍译成主语，而原句中的被动意义用"被""受到……""使……"等词来表达。

［例 5］ The model equation is reconciled by mathematical calculation with the actual situation.

通过数学计算使模型方程符合实际情况。

［例 6］ Durability is greatly influenced by concrete permeability.

混凝土的耐久性受其渗透性影响非常大。

［例 7］ The compressive strength of concrete is controlled by the amount of cement, aggregates, water, and various admixtures contained in the mix.

混凝土的抗压强度为水泥、集料、水及混合料中所含的各种添加剂的用量所控制。

(3) 把原句中的被动语态谓语动词分离出来，译成一个独立结构

［例 8］ It is believed that the automobile is blamed for such problems as urban area expansion and wasteful land use, congestion and slum conditions in the central areas, and air and noise pollution.

人们（有人）认为汽车造成一系列问题，如城市膨胀、土地浪费、市区拥挤、油污遍地以及空气和噪声污染等。

这种方法常用于一些固定句型中，类似的结构还有：

It is asserted that...	人们（有人）主张……
It is suggested that.	人们（有人）建议……
It is stressed that...	人们（有人）强调说……
It is generally considered that...	大家认为……
It is told that...	人们（有人）曾经说……
It is well known that...	众所周知……

有时，某些固定句型翻译时不加主语，如：

It is hoped that...	希望……
It is supposed that...	据推测……
It is said that...	据说……
It must be admitted that...	必须承认……
It must be pointed out that...	必须指出……
It will be seen from this that...	由此可见……

2.6.2 Negative Sentence Pattern 否定句型

英语的否定句多种多样。与我们所熟知的一般否定形式不同的是，英语中有一些特殊的否定句，其否定形式与否定概念不是永远一致的，它们所表达的含义、逻辑等都和我们从字面上理解的有很大差别。总之，英语当中的否定问题是一种常见而又复杂的问题，值得特别的重视。

(1) 否定成分的转译

否定成分的转译是指由意义上的一般否定转为其他否定（特指否定），反之亦然。常见的句型如下。

①“not... so... as...”结构

在谓语否定的句子中，如果带有 so...as 连接的比较状语从句，或 as 连接的方式状语从句，就应该译成“不像……那样……”，而不能直译成“像……那样不……”。

［例 9］ The sun's rays do not warm the water so much as they do the land.

太阳光线使水增温不像它们使陆地增温那样高。

②“not... think/believe...”结构

表示对某一问题持否定见解的句子，要把英语里面谓语动词 think，believe 等后面的否定词 not 转译到后面，即译为“认为……不……”“觉得……不是……”。

［例 10］ Ordinarily we don't think air as having weight.

我们通常认为空气没有重量。

③“not... because...”结构

在汉译时要注意这种结构可以表示两种不同的否定含义，既可以否定谓语，也可以否定原因状语 because，因此汉译时要根据上下文意思来判断。

［例 11］ This version is not placed first because it is simple.

这个方案并不因为简单而放在首位。

这个方案因为太简单所以不能放在首位。

以上的两种翻译实际上都是可以的。但是如果在上面的例子后面再加上一句“We need a more particular one which could explain every specific steps we have to take care of.”那么，就只能选择第二种译法了。

(2) 否定语气的改变

英语中的否定句并非一概译成汉语的否定句，有些否定句表达的是肯定的意思，常见的是 nothing but 句型；有些在特定的语义环境下也表达肯定意思。

［例 12］ Early computer did nothing but compute：adding，subtraction，multiplying and dividing.

早期的计算机只能做加减乘除运算。

(3) 部分否定

英语中 all，both，every，each，always 等词与 not 搭配使用时，表示部分否定。一般译成“不是都”“不总是”“不全是”。

［例 13］ All these building materials are not good products.

这些建筑材料并不都是优质产品。(不能译成“所有这些建筑材料都不是优质产品”)

类似的结构还有：“not... many”(不多)，“not... much”(一些)，“not... often”(不经常)。应该说明的是，“all... not...”和“not all...”这两种表示部分否定的形式，前者是传统的说法，虽然不合逻辑，但习惯上使用；后者是新说法，从逻辑和语法上着眼，认为比较合理，所以越来越多的人采用的是后一种用法，尤其在美国书刊中更为常见。

(4) 意义否定

有些句子当中没有出现否定词，但句中含有表示否定的词或词组，那么汉译时一般要将其否定意义译出，成为汉语的否定句。

［例 14］ The analysis is too complicated us to complete the computation on time.

分析工作太复杂，难以按时完工。

［例 15］ He gained little advantage from the scheme.

他从这项计划中没有得到多少利益。

常见的含有否定意义的词组还有：

but for	如果没有
free from	没有，免于
short of	缺少
in vain	无效，徒劳
make light of	不把……当一回事
in the dark	一点也不知道
safe from	免于
far from	远非，一点也不
but that	要不是，若非
fail to	没有

(5) 双重否定

① 针对同一事物的否定

有些词句在形式上是否定，而意思却是双重否定，也就是两个否定，即语法否定（not）和语义否定（非 not 否定词的词或词组否定，如 unexpected），都是针对同一事物而言，是“否定的否定”。

［例 16］ With a careful study of all the preliminary data made available to this engineer，there could be nothing unexpected about the problem.

通过这位工程师对所有初测资料进行审慎研究，这个问题就一切都在意料之中了。

［例 17］ There is no material but will deform more or less under the action of force.

在力的作用下，没有一种材料不或多或少地发生变形。（but 作关系代词，相当于 that... not）

常见的搭配还有：“not... until”（直到……之后，才能）“not（none）... the less”（并不……就不），“not a little”（大大地）。

② 针对两种不同事物的否定

两个否定分别针对两种不同事物而言，“不是否定的否定”，只是一句话里存在两个否定含义的词汇而已。

［例 18］ There is no steel not containing carbon.

没有不含碳的钢。

［例 19］ One body never exerts a force upon another without the second reacting against the first.

一个物体对另一个物体施作用力必然会受到另一个物体的反作用力。

2.6.3 Emphatical Sentence Pattern 强调句型

强调句型“It is（was）＋被强调部分＋that（which，who）...”几乎可用于强调任何一个陈述句的主语、宾语或状语。需要注意的是，此强调句型与带形式主语 it 的主语从句很相似，但它与主语从句不同的是，去掉以上几个英文单词以后，强调句中剩下的单词仍能组成一个完整的句子。

［例 20］ It is these drawbacks which need to be eliminated and which have led to the search for new methods of construction.

正因为有这些缺点需要消除，才导致了对施工新方法的研究探求。

［例 21］ It is this kind of steel that the construction worksite needs most urgently.

建筑工地最急需的正是这种钢。

"It is (was) not until＋时间状语＋that..."是强调时间状语常见的一种句型，可译成"直到……才……"。

［例 22］ It is not until 1936 that a great new bridge was built across the Forth at Kincardine.

直到 1936 年才在肯卡丁建成一座横跨海口的新大桥。

在强调句中，被强调的部分不仅可以是一个词或词组，而且还可以是一个状语从句。

［例 23］ It is not until the stiff concrete can be placed and vibrated properly to obtain the designed strength in the field that the high permissible compressive stress in concrete can be utilized.

只有做到在工地正确灌注振捣干硬性混凝土并使之达到设计强度时，才能充分利用混凝土容许压应力。

2.7 Translation of Long Sentence 长句的翻译

长句一般都是含有几个错综复杂关系的主从复句或并列复句，少数则是难用汉语表达的简单句。英语长句的理解，关键在于语法分析。具体来说，理解长句大体可以分为两个步骤进行：

① 判断出句子是简单句、并列句，还是主从句；

② 先找出句中的主要成分，即主语和谓语动词，然后再分清句中的宾语、状语、表语、宾语补足语、定语等。

英语长句的翻译主要采用分句和改变语序的方法，具体包括顺译法、倒译法和拆译法等。

2.7.1 Methods of Translation in Order Sentence 顺译法

对专业英语而言，只要不太违反汉语的行文习惯和表达方式，一般应尽量采用顺译。顺译有两个长处：一是可以基本保留英语语序，避免漏译，力求在内容和形式两方面贴近原文；二是可以顺应长短句相替、单复句相间的汉语句法修辞原则。

(1) 在主谓连接处切断

［例 1］ The main problem in the design of the foundations of a multi-storey building under while the soil settles is to keep the total settlement of the building within reasonable limits, but specially to see that the relative settlement from one column to the next is not great.

在土壤沉降处设计多层建筑基础的主要问题，就是要使建筑物的总沉降量保持在合理的限度内，而且特别要注意相邻柱子之间的相对沉降量不能过大。

(2) 在并列或转折连接处切断

［例 2］ Anything that can be done to reduce congestion and allow people to travel to the town center in a shorter time, will make the central area more accessible and, thus, will help people to decide to shop there as against in the suburbs or out of town.

为了减少拥挤，并能使人们用较短的时间到达中心而采取的任何措施将会使市中心区成为人们愿意去的地方，从而有助于人们决定到市中心购物，而不到郊区或出城去购物。

［例 3］ Park-and-ride differs from park-and-walk not only in the fact that the car park is much farther away from the town center, but also in that its success is much more dependent on the voluntary cooperation of the motoring public.

停车改乘与停车改步行的不同点不仅在于其停车场离市中心远得多，还在于这种停车方式的成败更多地取决于驾车进城的人们能否自愿合作。

(3) 在从句前切断

［例 4］ In the course of designing a structure, you have to take into consideration what kind of load structure will be subjected to, where on the structure the said load will do what is expected and whether the load on the structure is applied suddenly or gradually.

在设计结构时必须考虑到：结构将承受什么样的荷载，荷载作用在结构的什么位置，起什么作用，以及这荷载是突然施加的，还是逐渐施加的。

2.7.2 Methods of Translation in Inverted Sentence 倒译法

在英译汉时，常常需根据汉语的行文习惯表达方式将英语长句进行全部倒置或局部倒置。当然，翻译时只要能做到顺译，就不一定非要倒译，在大多数情况下，倒置也只是一种变通手段，并不是唯一可行的办法。

(1) 将英语原句全部倒置

［例 5］ About one third of all accidents happen when it is dark although obviously there is more traffic during daytime.

虽然在白天交通运输显然繁忙得多，但大约 1/3 的事故发生在晚上。

(2) 将英语原句部分倒置（将句首或首句置于全句之尾）

［例 6］ It is most important that the specifications should describe every construction item which enters into the contract, the materials to be used and the tests must meet, methods of constructions in particular situations, the method of measurement of each item and the basis on which payment should be calculated.

对于合同所列的各项施工项目、需要的材料及其检验要求、具体条件下的施工方法、每个施工项目的验收方法以及付款计算的依据等，说明书中都应加以详细说明，这一点是十分重要的。

2.7.3 Methods of Translation in Taking Apart 拆译法

为汉语行文方便，有时将英文原文的某一短语或从句先行单独译出，并利用适当的概括性词语或通过一定的语法手段把它同主语联系在一起，进行重新组织。

［例 7］ The loads a structure is subjected to are divided into dead loads, which include the weights of all the parts of the structure, and live loads, which are due to the weights of people, moveable equipment etc..

结构物受到的荷载分为恒载和活载，恒载包括该结构各个部分的重量，活载是由人的重量、可移动设备的重量等所引起的。

［例 8］ The integrated products quality control system used by thousands of enterprises in Russia is a combination of controlling bodies and objects under control interacting with the help of material, technical and information facilities when exercising QC at the level of an enterprise.

俄罗斯成千上万家企业采用的产品质量综合管理体系，是通过在整个企业范围内实行质量管理、把企业内各个管理机构和各种管理对象联结一起的综合体，这种联结是借助于材料部门、技术部门和信息部门实现的。

2.8 Translation of Subordinate Clause 从句的翻译

英语句子的某些成分由句子代替了单词以后，就形成了主从复合句。英语的主从复合句按语法功能来分有主语从句、宾语从句、定语从句、状语从句、表语从句和同谓语从句。主从复合句由于其结构相对简单句复杂，并且从句相对扩充以后能使复合句的结构更加复杂，因而它往往也是英语翻译时一个必须重视的问题。

2.8.1 Subject Clause and Object Clause 主语从句和宾语从句

在专业英语中，较常使用带形式主语 it 的主语从句，即句子常用引导词 it 作形式主语并放在句首，而把从句（真正的主语）放在谓语之后。

(1) 译成宾语从句

［例 1］ It is generally accepted that fatigue strength is drastically lower if the concrete is cracked.

人们普遍认为，混凝土若出现裂缝，其疲劳强度就会大大降低。

(2) 译成并列分句

［例 2］ It remains to be confirmed that epoxy coatings will retain their integrity over long periods of time in alkaline environments.

长期处于碱性环境中的环氧涂层能否保持其完好无损的性能，这有待进一步研究证实。

(3) 谓语分句

［例 3］ It is a fact that no structural material is perfectly elastic.

事实上，没有一种结构材料是完全的弹性体。

(4) 宾语从句

常见的形式宾语句的真实宾语也有三种：从句、不定式或动名词。形式宾语 it 和后面的说明语（多为形容词）在逻辑上是主表关系。它的翻译方法和形式主语句类同。

2.8.2 Attributive Clause 定语从句

定语从句在英语中的应用极广。由于英语中定语从句有长有短，结构有简有繁，对先行词的限制有强有弱，翻译时就不能一概对待，必须根据每个句子的特点，结合上下文灵活处理。一般来说，定语从句在逻辑意义上往往与所限定的词有着表示“目的”“结果”“原因”“让步”等含义。因此，在英译汉时，需要先弄清定语从句与先行词的逻辑关系。

(1) 译成前置定语

限制性定语从句往往译成前置定语结构，即译成“……的”。但有些非限定性定语从句有时也可以作前置处理，尤其是当从句本身较短，或与被修饰词关系较为密切，或因拆译造成译文结构松散时。

［例 4］ A drainage blanket is a layer of material that has a very high coefficient of permeability.

排水层为渗透系数较大的材料层。

［例 5］ In the design of concrete structures, an engineer can specify the type of material that he will use.

在混凝土结构设计中，工程师可以指定他将要使用的材料品种。

(2) 译成谓语

当关系代词在定语从句中充当主语且句子的重点是在从句上时，可省去关系代词，而将定语从句的其余部分译为谓语结构，以先行词充当它的主语，从而使先行词与定语从句合译成一句。

[例 6] A code is a set of specifications and standards that control important technical specifications of design and construction.

一套规范和标准可以控制设计和施工的许多重要技术细节。

(3) 译成并列句

非限制性定语从句往往需要拆译成并列句，有时，限制性定语从句因从句本身太长，前置会使句子显得臃肿，故也可采用拆译分列。

[例 7] The tendons are frequently passed through continuous channels formed by metal or plastic ducts, which are positioned securely in the forms before the concrete is cast.

预应力钢筋束穿入用金属管或塑料管制成的连续孔道，而金属管或塑料管在浇筑混凝土之前被固定在模板之中。

(4) 译成状语从句

定语从句有时与主语之间的关系，实际上是原因、条件、目的、让步、结果、转折等隐含逻辑关系。因此，英译汉时应以逻辑为基础，以忠实表达原文的意思为前提，将定语从句转译成汉语的状语从句。

[例 8] This is particularly important in fine-grained soils where the water can be sucked up near the surface by capillary attraction.

在细颗粒土壤中这一点尤其重要，因为在这种土壤中，由于毛细作用，水能被吸引到靠近道路表面的地方。(译成原因状语从句)

(5) 译成单句中的一部分

限制性定语从句有时在翻译时可压缩成宾语、谓语、表语和同位语。

[例 9] Fig. 1 incorporates many of the factors which must be considered in developing a satisfactory system.

图 1 所示的许多因素，在研制一个性能良好的系统时必须予以考虑。

2.8.3 Adverbial Clause 状语从句

状语从句相对而言比较简单，但有几点关于时间状语从句和地点状语从句的情况值得注意。

(1) 时间状语从句

时间状语从句在英语句中的位置相对灵活，但汉译时，有时候就要注意它们的位置问题。汉语习惯是先发生的事情先讲，表示时间的从句汉译时要提前。当时间顺序很明显时，有时还可以省略关系副词。

[例 10] Pre-tensioning is a method of prestressing in which the steel tendons are tensioned before the concrete has been placed in the moulds.

先张法是一种在往模板内浇筑混凝土之前，即将钢筋束张拉而施加预应力的方法。

值得注意的是，有时看似由 when，while 等引导的时间状语从句，实际上却是具有条件状语从句或是让步状语从句的意义，即相当于 if/although 引导的状语从句，翻译时往往可以转译成条件状语从句。

[例 11] On the site when further information becomes available, the engineer can make changes in his sections and layout, but the drawing office work will not have been lost.

在现场若能取得更确切的资料，工程师就可以修改他所做的断面图和设计图，但是绘图室的工作并非徒劳无功。

(2) 地点状语从句

由 where 引导的状语从句，有时不宜译作地点状语从句，因为原文实际上所表达的不是地点意义而是条件意义，状语从句起着条件状语的作用。因此，若遇到这类地点状语从句时，一般可以转译成条件状语从句。

[例 12] Where internal corrosion is known to exist, the following practices can be employed.

如果发现有内腐蚀的存在，可采用以下措施。

2.9 Translation about Quantity 有关数量的翻译

2.9.1 Doubled and Redoubled Addition 成倍增加

(1) 表示数量成倍增加的句型

基本句型有以下几种：

A is *N* times are large (long, heavy, ...) as *B*.

A is *N* times larger (longer, heavier, ...) than *B*.

A is larger (longer, heavier, ...) than *B* by *N* times.

上述几个句型的含义相同，均可译成：*A* 的大小（长度、重量……）是 *B* 的 *N* 倍，或 *A* 比 *B* 大（长、重……）*N*－1 倍。

[例 1] The temperature on the site may be to 40 times higher in summer as compared to winter.

工地的夏季气温可能是冬季气温的 40 倍。

(2) 表示倍数的单词

有些单词可直接表示倍数关系，如 double（增加 1 倍，翻 1 番），treble（增加 2 倍，或增加到 3 倍），quadruple（增加 3 倍，翻 2 番）等。

[例 2] If the speed is doubled, keeping the radius constant, the centripetal force becomes four times as great.

若保持半径不变，速度增大 1 倍，则向心力增大为原来的 4 倍。(即增大了3 倍)

还有些表示增加的动词（如 increase）加上 *N* times，by *N* times，*N*-fold 等来表达“增加 *N* 倍”的含义。

[例 3] Such construction procedure can increase productivity over three-fold.

这种施工工序可使生产率提高到 3 倍以上。(即提高了 2 倍多)

2.9.2 Doubled and Redoubled Reduction 成倍减少

语句中表示成倍减少含义时，通常包含以下句子成分：

reduce by *N* times

reduce *N* times

reduce to *N* times

N-fold reduction
reduce *N* times as much (many...) as
reduce by a factor of *N*
reduce *N*-fold
N times less than
对上述结构，均可译成“减少了（*N*－1)/*N*”或“减少到原来的 1/*N*”。
［例 4］ The production cost has reduced four times.
生产成本减少了 3/4。(即减少到原来的 1/4)
［例 5］ The advantage of the present scheme lies in a fivefold reduction in manpower.
这一方案的优点在于节约人工 4/5。(即节约到原来的 1/5)

2.9.3 Uncertain Quantity 不确定数量

英语中常用来修饰不确定数量的词有：circa，about，around，some，nearly，roughly，approximately，or so，more or less，in the vicinity of，in the neighborhood of，a matter of，of the order of 等。这些词可译成“大约……”“接近……”“……上下”“……左右”等。

［例 6］	a weight around 12 tons	12 吨左右的重量
	300km or so	大约 300 千米
	a force of the order of 100kN	约为 100 千牛的力

另外，还有一些表示不确定数量量级的词组，如：

teens of...	十几（13～19）
decades of...	几十
scores of...	几十（多于 40）
thousands of...	几千
tens of...	几十
dozens of...	几打
hundreds of...	几百

另外，度量单位的英语表达及换算可参见 Appendix。

Exercises

Translate the following into Chinese.

1. Rock made under water tell another story.
2. Force is any push or pull that tends to produce or prevent motion.
3. A prediction of the duration of the period when building materials cannot be supplied would be of value in the planning of construction.
4. Our present-day civilization could never have evolved without the skills included in the field of engineering.
5. In a large structure expansion joints are always provided so that the material may be allowed to expand.
6. Bituminous seals are placed on the joints between concrete slabs to prevent the ingress of water.
7. This paper aims at discussing the properties of the newly discovered material.
8. The combination of mechanical properties of this alloy can be well achieved by heat treatment.
9. The Congressman tends to be very interested in public works——such as a new government buildings, water projects, highways and bridges, etc. ——that will bring money to the area or improve living conditions.
10. For any unusual structure the tasks of design and analysis will have to be repeated many times until, after many calculations, a design has been found that is strong, stable and lasting.

Unit 3 Writing of Scientific and Technical Papers

撰写英文科技论文的目的是为了参与国际间学术交流，如在英文期刊杂志上发表或在国际学术会议上宣读自己的科技论文，让同行了解和分享你的学术成果。

为提高论文写作质量、减少撰写过程中的盲目性，有必要较系统地了解和学习英文科技论文的写作方法。本章结合土木工程，介绍英文科技论文写作的一般方法，并通过实例解释写作要点和技巧。

3.1 Stylistic Rules of Papers 论文体例

国际标准化组织（International Organization for Standardization）、美国国家标准化协会（American National Standards Institute）和英国标准协会（British Standards Institute）等国际组织都对科技论文的写作体例（stylistic rules）做出了规定，其基本内容如下。

3.1.1 Composition about Papers in Periodical 期刊类论文组成

对于期刊类科技论文，主要部分包括以下几点。

- Title 标题
- Abstract 摘要
- Keywords 关键词，或主题词（Subjects）
- Main text 正文

  Introduction 引言

  Analysis of the theory，test procedure 理论分析或试验过程

  Results 结果

  Discussions（Summary，Conclusions，Suggestion and Development）讨论（总结，结论，建议和发展）
- Acknowledgments 致谢（可能没有）
- References（Appendix）参考文献（附录）

3.1.2 Composition about Science Report 长篇科学报告组成

长篇科学报告包括科研成果、学位论文、可行性研究等。

(1) Front 前部

- Front cover 封面

  The title 标题

  Contract or job number 合同或任务号

  The author or authors 作者

  Date of issue 完成日期

Report number and serial number 报告编写和系列编号

Name of organization responsible for the report 研究单位名称

A classification notice 密级

- Title page 扉页
- Letter of transmittal（Forwarding letter）提交报告书
- Distribution list 分发范围
- Preface or foreword 序或前言
- Acknowledgments 致谢（可能没有）
- Abstract 摘要
- Table of contents 目录
- List of illustration 图表目录

(2) Main Text 正文

- Introduction 引言
- Analysis of the theory，test procedure 理论分析或试验过程
- Results 结果
- Discussions（Summary，Conclusions，Suggestion and Development）讨论（总结，结论，建议和发展）

(3) Back 后部

- References 参考文献
- Appendix 附录
- Tables 表
- Graphics 图
- List of abbreviations，signs and symbols 缩写、记号和符号表
- Index 索引
- Back cover 封底

实际写作中，不一定也不可能完全按上述内容编写，视具体情况和要求确定。

3.2 Title and Sign 标题与署名

论文标题属于特殊文体，一般不采用句子，而是采用名词、名词词组或名词短语的形式，通常省略冠词。从内容上，要求论文标题能突出地、明确地反映出论文主题。具体而言，在拟定论文标题时应注意以下几点：

① 恰如其分而又不过于笼统地表现论文的主题和内涵；

② 单词的选择要规范化，要便于二次文献编制题录、索引、关键词等；

③ 尽量使用名词性短语，字数控制在两行之内。

［例 1］ Bayesian Technique for Evaluation of Material Strengths in Existing Structures

采用贝叶斯技术评估既有结构的材料强度

3.2.1 Normal Format of Writing Title 标题书写的几种常用格式

(1) 标题主要单词首字母大写，其余小写

［例 2］ Bridge Live-Load Models

(2) 标题主要单词首字母大写，其余为小型大写

［例 3］ NONLINEAR ANALYSIS OF SPACE TRUSSES

(3) 标题文字全部大写

［例 4］ RELIABILITY ASSESSMENT OF PRESTRESSED CONCRETE BEAMS

(4) 标题首单词首字母大写

［例 5］ Sustainable development slowed down by bad construction practices and natural and technological disasters

3.2.2 Sign and Information of Author 署名和有关作者信息

一般紧跟在论文标题之后的是论文署名和有关作者的信息，如作者单位、通信地址（近年来还包括 E-mail 地址、个人主页的网址）、职称、学衔或会员情况等。按照英语国家的习惯，论文署名时名在前（可缩写），姓在后；但为了便于计算机检索，也有姓在前、名在后的情况（参考文献中的作者姓名排列就是这样）。有关作者的信息有时放在署名之后，有时放在论文第一页的页脚，有时放在论文的末尾，有时还分开编排，这要视论文载体的具体要求而定。

［例 6］ 作者信息紧接在署名之后。

Developing Expert Systems for Structural Diagnostics and Reliability
Assessment at J. R. C
A. C. Lucia
Commission of the European Communities，Joint Research Center，
ISPRA Establishment，21020 ISPRA（VA），Italy

［例 7］ 作者信息放在论文第一页的页脚。

BRIDGE RELIABILITY EVALUATION USING LOAD TESTS
By Andrzej S. Nowak[1] and T. Tharmabala[2]

在论文第一页的页脚：

[1]Assoc. Prof. of Civ. Engrg.，Univ. of Michigan，Ann Arbor，MI 48109

[2]Res. Ofcr.，Ministry of Transp. and Communications，Downsview，Ontario，Canada M3M 1J8

注意，在作者信息以及参考文献内，为节省篇幅，会采用较多的甚至不常见的缩写。

如例 7 中的 Assoc 为 Associate，Civ 为 Civil，Engrg 为 Engineering，MI 为 Michigan，Res 为 Research，Ofcr 为 Officer，Transp 为 Transportation 等。

3.3 Abstract 摘要

摘要（abstract）是一篇科技论文的核心体现，直接影响读者对论文的第一印象。一篇学术价值较高的论文，若摘要撰写得不理想，会使论文价值大打折扣。因此，掌握英文摘要的特点是非常重要的。

3.3.1 The Basic Characters of Abstract 摘要的基本特点

① 能使读者理解全文的基本要素，能脱离原文而独立存在。

② 摘要是对原文的精华提炼和高度概括，信息量大。

③ 具有客观性和准确性。

3.3.2 The Form and Requirement of Contents 形式和内容要求

摘要和基本形式和内容表现在以下几方面。

① 若无特殊的规定，一般摘要位于论文标题和正文之间，但有时也要求接在正文之后。

② 对于一般篇幅的论文，摘要的篇幅控制在 80～100 单词左右；对于长篇报告或学位论文，摘要的篇幅控制在 250 单词左右，一般不超过 500 个单词。

③ 一般篇幅论文摘要不宜分段，长篇报告或学位论文的摘要可分段，但段落不宜太多。

④ 与标题写作相反，摘要需采用完整的句子，不能使用短语；另外，要注意使用一些转折词连接前后语句，避免行文过于干涩单调。

⑤ 避免使用大多数人暂时还不熟悉或容易引起误解的单词缩写和符号等；不可避免时，应对这些单词缩写和符号在摘要中第一次出现处加以说明；例如：TM（Technical Manual）、CCES（Chinese Civil Engineering Society ）等。

⑥ 摘要的句型少用或不用第一人称，多采用第三人称被动语态，以体现客观性。

⑦ 避免隐晦和模糊，采用准确、简洁的语句概括全文所描述的目的、意义、观点、方法和结论等。

⑧ 注意体现摘要的独立性和完整性，使读者在不参看原文的情况下就能基本了解论文的内容；摘要的观点和结论必须与原文一致，忌讳把原文没有的内容写入摘要。

⑨ 通常摘要采用一个主题句（topic sentence）开头，以阐明论文的主旨，或引出论文的研究对象，或铺垫论文的工作等，避免主题句与论文标题的完全或基本重复。

⑩ 在摘要之后，通常要附上若干个表示全文内容的关键词 ，或主题词，或检索词（indexing term），应选用规范化的、普遍认可的单词、词组或术语作为关键词，不宜随心编造。

3.3.3 The Sentences Pattern in Common Use 常用句型

在撰写摘要时，可套用一些固定句型。不过，掌握句型和词汇特点，并结合实际情况灵活运用更为重要。下列几个句型仅供参考。

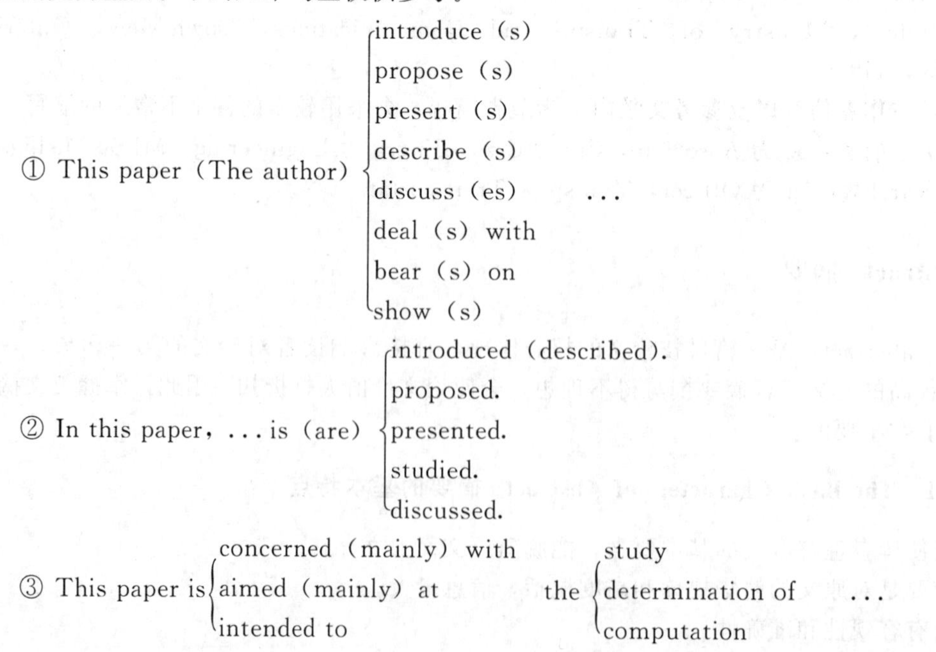
① This paper（The author） { introduce（s） / propose（s） / present（s） / describe（s） / discuss（es） / deal（s） with / bear（s） on / show（s） } ...

② In this paper，... is（are） { introduced（described）. / proposed. / presented. / studied. / discussed. }

③ This paper is { concerned（mainly） with / aimed（mainly） at / intended to } the { study / determination of / computation } ...

④ The {chief aim / main purpose / primary object / primary objective} of {the present study is / this investigation was / our research has been / these studies will be} to {obtain some / review the process / assess the role / find out what reveal the cause / of establish the equation} …

⑤ … {has (have) / has (have) be} {concluded / gained / obtained / yielded / arrived at / generated / acquired / achieved} …

⑥ {These studies / The research} lead(s) {us / the author(s) / the writer(s)} to {conclude / suggest / postulate that / a conclusion / a belief} …

［例 1］ 本文阐述了建设“信息高速公路”的目标、内容、意义及其影响。

This paper describes the objects, contents, significance and impact of Information Super Highway project being constructed.

［例 2］ 本文的主旨为：采用更为合理的系统可靠度理论，促进结构设计和评估规范的发展。

The main purpose of this paper is to contribute to the development of more rational system reliability-based structural design and evaluation specifications.

［例 3］ 该方法基于极坐标空间分割以及单位中线向量的自动生成技术。

The method is based on a radial-space division technique in conjunction with automatic generation of unit centerline vectors.

［例 4］ 数值算例的结果表明，与蒙特卡罗模拟方法相比，建议方法的精度较高。

Results of numerical examples indicate that the proposed method has good accuracy with the Monte-Carlo simulation method.

［例 5］ 建议的方法可作为分析混凝土箱梁畸变应力的基础。

The proposed approach may be used as a basic for the analysis of distortion-induced stresses in the concrete box girders.

3.3.4 The Example of Abstract 摘要实例

这是一篇研究组合桥极限荷载的文章摘要，标题是：Ultimate Loads of Continuous Composite Bridges（连续组合桥的极限荷载）。

Abstract：In this paper, the prediction of the most probable yield-line patterns of failure for relatively wide composite bridges is presented. The prediction is based on a parametric study as well as on laboratory test results on composite bridge models. The degree of fixity between the transverse steel diaphragms and the longitudinal steel girders is considered with respect to influence on the ultimate load capacity of the bridge. Good agreement is shown between the theoretical and experimental results. A method of relating AASHTO truck loading to the collapse load is presented. The derived equations can be used either to predict the ultimate load capacity or the required ultimate moments of resistance for design of simple-span and of continuous-span composite bridges.

参考译文

摘要：本文针对相对较宽的组合桥，基于参数研究以及实验室模型试验结果，预测出最可能出现的失效屈服线模式。对钢横隔板与纵向主梁的连接紧固程度也加以考虑，以计及其对桥梁极限承载力的影响。实验结构与理论分析结构很接近。(文中) 提出了一种把AASHTO 车辆荷载与破坏荷载相关联的方法。所推导的方程可用来预测极限承载力，或用来设计简支或连续组合桥的极限抵抗弯矩。

3.4 Writing and Organizing of the Main Text 正文的组织与写作

正文是论文的主体部分。由于学科、论题、方法和手段的差异，正文的组织和写作也不可能千篇一律。总的原则应该是：论文的结构层次分明，逻辑关系清晰，研究重点突出，语言文字简约。通常，正文包括以下几部分。

① 介绍与论题相关的背景情况和研究现状，并提出问题。

② 对理论分析过程、应用材料、计算方法、应用软件、实验设备、研究过程等的详细描述。

③ 对计算、分析或实验研究结果进行分析讨论，提出结论和建议及发展方向等。

3.4.1 The Grammar in Common Use 常用语法

常用语法主要包括一般现在时、现在完成时、无人称被动语态、条件语句、祈使语句等。

(1) 一般现在时和现在完成时

在科技类英语的写作中，一般现在时（包括被动语态）用得最多，常用来描述不受时间限制的客观事实和真理，表达主语的能力、性质、状态和特征等。用得较多的还有现在完成时，但主要是被动语态，主要用来表述过去发生的（无确切时间），或在过去发生而延续到现在的事件对目前情况的影响。常与现在完成时连用的词有：already，(not) yet，for，since，just，recently，lately 等。

[例 1] 通过这样做，非破坏性试验的结果会提高极限强度结论的可靠度。

By so doing, the results of the nondestructive tests increase the confidence level of the ultimate strength results.

[例 2] 至今，多数的研究工作都是关于非破坏结构的，没有涉及其组成构件已经破坏的结构。

Most of the research works to date has been related to undamaged structures, and not to the structure as already by damage to its components.

另外，在论文的引言部分论述某一研究课题的过去情况和目前进展时，时常会用到其他时态，如过去时、现在进行时等。

(2) 无人称被动语态

对无需说明或难以说明动作发出者的情况，就用无人称被动语态。

［例 3］ 这点如图 1 所示。

This is shown in Fig. 1.

［例 4］ 材料的性能、尺寸和分析模型的准确性被当作随机变量。

Material properties, dimensions, and accuracy of the analysis model are treated as random variables.

(3) 条件语句

在理论描述中，常常用到一些条件语句，说明一种假设情况。最常用的条件语句为 if 语句。此外还有其他一些条件表达方式，如：unless（=if... not）、providing（that）、provided（that）、only if、given+名词、in case、so（as）long as、suppose（that）、assume（that）、with... 等。

［例 5］ 因此，如果知道产生屈服的应力组合，应变矢量的方向可以根据流动定律被唯一确定。

Hence, if the combination of stresses which cause yield are known, the direction of the strain vector can be determined uniquely from the flow rule.

［例 6］ 如果给出风速和环境条件，就有可能预测风对建筑的作用。

Given wind speed and environmental conditions, it is possible to predict the actions by wind on buildings.

［例 7］ 只要知道荷载条件，作用在结构构件上的力就很容易被分析。

Provided that the load conditions are known, the forces on structural members can be analyzed easily.

［例 8］ 如果有设备，试验将很容易实现。

With the equipment the experiment would be readily conducted.

［例 9］ 除非特殊说明，这篇论文中 L 将代表梁长。

L will represent alone the length of the beam in the paper unless otherwise stated.

［例 10］ 只有当 $x<0$ 时，这些方程才成立。

These equations will hold as long as $x<0$.

(4) 祈使语句

在理论解释、公式说明和试验分析中，经常会用到祈使语句。它表示指示、说明、建议，或表示条件、假设、设想等。

［例 11］ 值得注意的是，混凝土是多孔的材料，空气中的二氧化碳能渗入到混凝土内部。

Note that concrete is a porous material and the carbon dioxide from the air can penetrate into the interior of it.

［例 12］ 在方程（1）中，令 a 等于 b。

Let a equal b in the Equation (1).

［例 13］ 确定模板固定在正确的位置。

Be sure to fix the mould board in right position.

［例 14］ 假如忽略温度的影响，方程可以被写成如下形式。

Suppose that the influence of temperature is negligible the equation cam be rewritten as follows.

3.4.2 The Sentence Pattern in Common Use 常用句型

一般在科技论文撰写中采用符合语法的任何句型，并无一定之规。大量采用的仍是“主＋谓＋宾”和“主＋系＋表”结构及复合句型等。不过，有些句型简单明了、适应性广，使用频率较高。现列举几种句型如下。

① It {系动词＋形容词 / 动词} ＋that 从句

② It＋系动词＋形容词＋to

主要形式有：

It {is / was / seems / appears / becomes / proves / usual（unusual）} ＋ {easy（difficult） / possible（impossible） / necessary（unnecessary） / useful（useless） / practical（impractical） / common（uncommon） / important（unimportant）} ＋to（动词不定式短语）

③ 主语＋系动词＋to（动词不定式短语）

［例 15］ 人们建议防水膜应该覆盖女儿墙间的整个顶面。

It is recommended that waterproof membranes should cover the entire deck surface between parapets.

［例 16］ 在分析试验结果时使用无量纲的参数是方便的。

It is convenient to use non dimensional parameters in the analysis of the test results.

［例 17］ 这篇文章的目的是提出一个计算铁路路床荷载效应的实际程序。

The objective of this paper is to present a practical procedure to calculate the load responses for railway roadbed.

3.4.3 The Forms of Omitting 省略形式

专业英语中省略的情况较多，下面只列举几种较常见形式。

（1）用分词独立结构代替从句

当用主句和从句的主语相同时，从句的主语可以省略，用分词独立结构代替从句。句型为：从属连词（Before，After，When，While，On，By，In 等）＋分词，主语……

［例 18］ 在检测应力之前，应该首先校核试验设备。

Before measuring the stresses，the testing equipment should be checked first.（＝Before the testing equipment measures...）

［例 19］ 当使用平衡梁结构时，这个方法效果比较好。

When used with the balanced cantilever construction，the approach works best.（＝When the approach is used with...）

（2）用过去分词作后置定语代替定语从句

英语中常用 which，where，what，that 等引导后置定语从句，修饰前面的名词。当后置定语从句中的动词为被动时态时，可省略引导部分，直接用过去分词作后置定语，使句子更简练。下面几个例句括号中为省略部分。

［例 20］ 关于这个主题进行的研究是广泛的。

The research（which is）being carried on this subject is extensive.

［例 21］ 目前，用于模拟悬索桥施工整个过程的几种方法是可行的。

Currently，several methods（which are）used for simulating the entire process of construction of cable-stayed bridges are available.

（3）并列复合句的名词成分的省略

在并列复合句子中，其第二分句（或后续分句）里常省略与第一分句相同的句子成分（主语、谓语、宾语或状语），见下两例的括号内部分。

［例 22］ 如果梁向下弯曲，弯矩是正的。如果梁向上弯曲，弯矩是负的。

The bending moment is positive if the beam bends downwards.（the bending moment）negative if（the beam bends ）upwards.

［例 23］ 在图 2 中，*R* 代表抗力，*S* 代表荷载效应，*K* 代表安全系数。

In Fig. 2，*R* is the resistance，*S*（is）load effect and *K*（is）the safety factor.

（4）状语从句中句子成分的省略

在表示时间、地点、条件、让步、方式的状语从句中，若其主语与主句的主语一致且谓语含有动词 be，或其主语是 it，就可省略从句中的主语和作助动词或者连系动词的 be。

［例 24］ 钢梁遇热膨胀，遇冷收缩。

Steel girders expand when（they are）heated and contract when（they are）cooled.

［例 25］ 如所预计的，洪水给基础造成了严重的影响。

The flood scored the foundation seriously as（it had been）expected.

3.4.4 The Example Sentence and Explanation About Writing 写作例句与说明

写作情况多种多样，可采用的句型也不少。本节根据具体写作对象的不同，介绍一些语句结构、短语和词汇供参考。读者应结合实际情况适当选择并灵活运用，切忌死套。

（1）Progress and Commentary 进展与评述

在科技论文中，尤其在引言部分，往往首先需要对目前进展和他人工作进行评述。对这种情况，通常采用现在完成时态。若干语句结构如下。

① A substantial review of... has been given by...

② An extensive list of references can be found in the review paper by...

③ ... have attracted researchers' attention since...

④ There has been theoretical interest in the field of... for the last decade.

⑤ ... have been a major concern in the development of...

⑥ Recently this topic has seen tremendous growth in the theory and methods of...

⑦ Much progress has been made in...

⑧ The last decade has seen tremendous growth in the theory and methods of...

⑨ However，attention was just focused on...，not on...

⑩ Since... have been described in detail elsewhere，only a brief outline of the important aspects of... is presented here.

⑪ ... is far from simple and it is therefore desirable to...

⑫ There is a growing need for...

⑬ The problems of... are issues which have increasing important in 1990's...

⑭ Some attempts have been made to apply... to...

⑮ It has been shown by... that... have a significant effect on...

⑯ However, it has been observed that...

［例 26］ 自从 20 世纪 70 年代起，高速公路隧道施工的整个过程的模拟已经吸引了研究者的注意，包括现场的分析和设计。

The simulation, including analysis and design on site, of entire process for highway tunnel construction has attracted researches attention since 1970's.

［例 27］ 最近，关于桥梁设计的动荷载已经引起了极大的兴趣，比如移动的车辆、地震和风。

Recently, there have been an interesting interest in and concern about dynamic loads, such as moving vehicles, earthquake and wind, for bridge design.

［例 28］ 然而，人们观察到先前的研究不能清楚地显示相互关系……因为缺乏检测数据。

However, it has been observed that the previous researches could not show clearly the relationship... because of lacking of measured data.

(2) Definition and Description 定义与描述

在理论分析和公式推导中，常需要对一个事物或概念做出定义，并进行解释和描述。常见语句结构有：

① Define... to be...

② ... is { defined as / called / said to be }

③ ... { is / means / signifies / is considered to be / is taken to be / refers to }

［例 29］ 这里，*N* 被定义为在波动荷载作用下，在特定的应力水平下，使试件破坏需要的循环次数。

Here, *N* is defined as the number of circles required to break a specimen at a particular stress level under a fluctuating load.

［例 30］ β 被称为安全系数，用于测定所有类似的结构组成构件的安全程度。

β is called safety index, and is taken to be a measurement of safety level for all similar components of structures.

(3) Hypothesis and Assumption 假说与假设

假说（hypothesis）是在事实基础上根据类比推理、归纳推理和演绎推理提出的。假设（assumption）可用来预测事物发展趋势，简化分析和计算过程。常见语句结构有：

① { This hypothesis / The hypothesis of... } + { was developed / was put forward / was suggested } + { in 1980 / in the early 70's / in the late 70's }

② {This hypothesis that... / This hypothesis of...} + {has been checked / has not yet been verified remain aims / to be supported} + by...

③ {The theory is based on an assumption / These data lead us to assume / The author proceeds from an assumption} + that...

④ {What we obtained / What the authors report} + {(does not) contradict(s) / (does not) agree(s) with / is in agreement with / is not consistent with} + the assumption (that...)

[例 31] 这些数据令我们假设风产生的影响几乎是不重要的。

These data lead us to assume that it is of little importance to take the effect caused by wind into account.

[例 32] 我们在实验中获得的结果与理论假设相符得较好。

What we obtained in the test is in good agreement with the theoretical assumption.

(4) Classification and Comparison 分类与比较

分类与比较是根据事物的特点、属性进行归纳区别，并对两种以上同类事物的异同点或优缺点进行对比，以加深对事物本质的认识。分类语句结构如下。

① There are *N* {types / kinds (varieties) / classes / sorts} of...

② {may be / might be / can be} + {classified / divided / into... / categorized / grouped}

③ There are..., the first..., the second...

④ ...differs (is different) from...

比较语句结构有：

⑤ as + *adj.*/*adv.* 原形 + as... 和 not so (as) + *adj.*/*adv.* 原形 + as...

⑥ 倍数 + as...as...

⑦ *adj.*/*adv.* 比较级 + than...

⑧ ... {(be) superior to / (be) in advance of / (have) superiority over / (have) advantages over} (比……好，优于……)

⑨ ... {(be) inferior to / (be) nothing to} (比……差，不如)

⑩ ... {bear comparison to (with) / stand comparison to / (be) equal to / (be) comparable to} (比得上，与……相比)

⑪ 常用短语 { with respect to / as against / in comparison to / as compared to / compared to (with) / by the side of / in contract to } 相对于；与……相比；对比起来

［例 33］ 结构工程项目可以分为方案、设计和施工三个阶段。

A structural engineering project can be divided into three phases: planning, design and construction.

［例 34］ 通过比较，这个方案似乎优于前一个方案。

In comparison, the scheme seems superior to the previous one.

(5) Methods and Means 方法与方式

在阐述研究过程时，总要论及所采用的方法。在专业英语中，对方法的描述往往是句子的状语成分，内容涉及描述方法的类型、途径、意义、范围、方式等。

① 使用、采用（某方法）

by means of
by
with (by) the aid of
by virtue of
in terms of
by the use of
using

② 用（通过）……方法

mathematically	用数学方法
theoretically	通过理论探讨，理论上
statistically	用统计方法
empirically	用经验方法
experimentally	用实验方法

③ 以……方法

one way or another	以某种方法（式）
in a similar way	以类似方法
in all manner of ways	以各种方法
by some means or other	以某种方法
in much the same way	以基本相同的方法
in a regular manner	以常用方法
in the usual manner	以常用方法

④ 在……意义上

in a sense	在某种意义上
in all sense, in every sense	在各种意义上
in the same sense	在同样意义上
in this sense	在这个意义上
in a narrow sense	狭义上
in a broad sense	广义上

⑤ 各种方式

without break (intermission)	不间断地
intermittently, with intermittence	间断地
on and on, continuously	持续不断地
in fits and starts, on and off	断续地
in combination (conjunction) with	与……结合
in isolation	孤立地
independently	独立地
in a discrete fashion	以离散的方式
in an analogous manner	以类似的方式
after the manner (pattern, fashion) of	仿效，仿照
in chronological order	按时间顺序
in descending (ascending) order	以递降（升）顺序
clockwise, in a clockwise sense	顺时针地
counterclockwise	逆时针地
in groups	成群地
in a line	成直线地
in pairs	成对地
in rows	成排地
in a circle	成圆圈地
upside down	上端朝下
downside up	下端朝上
inside out	里面朝外
outside in	外面朝里
the right side up	正面朝上

［例 35］ 应变的变化可以通过应变仪或类似的仪器记录。

The change in strain can be recorded by means of strain gauge or an analogous apparatus.

［例 36］ 尝试以某种方式维护和改善现有建筑结构。

Attempts have been made to maintain and rehabilitate the existing building structures one way or another.

(6) Degree and Magnanimity 程度与度量

在论文写作中，需要对事物某一方面的水平进行修辞，以具体或不具体的方式进行度量。这部分内容通常作为句子的定语、表语或状语成分。可参考的词汇或词组如下。

① 显著的；用作定语和表语

pronounced, appreciable, noticeable, conspicuous, considerable, remarkable, marked, significant, substantial 等。

② 显著地，远……得多；作状语

significantly, substantially, considerably, a great deal, much, a lot, far 等。

③ 稍微；作定语或状语

a little, a bit, somewhat, slightly, more or less 等。

④ 大小

predominantly	主要（地）
to a less (slight) degree (extent)	在较小程度上

in a greater degree	在较大程度上
in a considerable degree	在很大程度上
to a high degree	在很高程度上
to a certain extent	在某种程度上
in some degree	在某种程度上

⑤ 粗细

at large	笼统地
in detail，in full length	详细地
in a more detailed fashion	更为详细地
in considerable detail	相当详细地
in more detail	比较详细地
in some detail	较为详细地

⑥ 范围

radically	根本上
on the whole	总的说来，大体上
in general，far and by	大体上，一般说来
essentially，basically，primarily，largely	基本上
main，chief，primary，major	主要的（定语）
mostly，mainly，chiefly，dominantly	主要的，大部分
predominantly，in the main，in a great measure	主要地，大部分
for the most part，for the greater part	主要地，大部分
entirely，completely，utterly，wholly，to the full extent	完全地
in part，partially，partly	部分地
extremely，in the extreme，to the last degree	极其，非常

⑦ 度量

on the order of	相当于……数量，大约
to within	…… 在……精度以内
by weight（volume，area...）	按重量（体积、面积）计量
a slope of...in（on，to）	……坡度……：……
a batter of...in	……斜度……：……
angle of...with	……与……成……角
at an angle of θ to	……与……成 θ 角
an a right angle to，squared to	与……成直角
in length（width，depth...）	按长度（宽度、深度……）
数字+单位+long（wide，deep...）	长（宽、深……）

［例 37］ 详细施工控制的研究一定程度上依赖于现场收集的数据（信息）。

The research of construction control in detail depends，in a considerable degree，on the data collecting from site.

［例 38］ 目前，缆绳中的拉力测定只能在±3%之内。

At present，the tensional force in cables can only be measured to within±3%.

［例 39］ 钢管为 30 米长，端部用 5 厘米厚、直径 30 厘米的铸铁圆盘封闭。

The steel tube is 30m long，and its ends are closed with a cast steel disk which is 5cm in thickness and 30cm in diameter.

(7) Scale and Rate (Proportion and Ratio) 比例与比率

在 Unit 2 中，曾论述过数的增减速、比较和倍数等表达，这里再介绍数的比例与比率的一般表达方式。常用语句结构如下。

① ……与……成比例

... (be) in the proportion of

② 与……成正比例

be a direct measure of

be a direct dependence upon(on)

vary (directly) as

vary in the direct ratio of

be (directly) proportional to

be in proportion to

be relative to

③ 与……成反比例

be in relation to

be an inverse measure of

be an inverse dependence upon(on)

vary inversely as

vary in the inverse ratio of

vary in the reciprocal of

be inversely proportional to

be in inverse proportion to

depend (s) inversely as

[例 40] 物体的加速度与作用力成正比，与物体的质量成反比。

The acceleration of a body is directly proportional to the force acting and is inversely proportional to the mass of the body.

[例 41] 对于在弹性范围内加载的低碳钢材，应力与应变成正比。

For mild steel loaded in elastic region, the stress varies directly as strain.

(8) Tables and Graph (Diagram) and Formula 图表与公式

在科技论文中，为了更加直观、简洁和明确地表述一定的概念、理论和应用，往往采用图表和公式。

① 图

与图有关的词汇有：graph，diagram，drawing，chart，sketch 等，如

curve line graph	曲线图
projection drawing	投影图
flow chart	流程图
diagrammatic sketch	示意图
key diagram	概略原理图
perspective drawing	透视图
histogram	直方图，频率曲线

在工程图纸中常用的词汇：plan（平面图）、side view（侧视图）、top view（俯视图）、elevation（立面图）、section（截面图）、detail（大样图）、scale（比例）等。

若论文较短，可将文中所有的图按顺序依次编号，如 Fig. 1，Fig. 2；对于较长的学位论文或报告，可分章节编号，如 Fig. 1-1，Fig. 1-2，或 Fig. 1. 1，Fig. 1. 2 等；图名跟在其后。另外，若采用的图引自其他文献，就需要在文中或图名后注明来源。

[例 42] 图 8.4——钢筋混凝土梁柱的典型构件强度模拟结果。

Fig. 8.4—Typical member strength simulation results for reinforced concrete beam-columns.

注明来源的方式有以下几种。

source：...	资料来源：……
photograph：...	图片取自……
furnished by permission of...	蒙……允许载用
courtesy of the...	蒙……特许刊用
from：... published by...	引用……出版的……
copyright...	本图版权为……所有

［例 43］ 如图 2 所示，竖向（纵）轴为应力（集度），水平（横）轴为应变。

As indicated in Fig. 2，stress intensity is shown on the vertical axis and strain on the horizontal.

在论文中，有时还需要对图中的符号及其位置等进行解释，举例如下。

at the top	在顶上
at the bottom right	在右下角
at the top left	在左上角
in the middle	在中间
upper（middle，lower）part	（一张图的）上（中、下）部
upper（lower）half	（一张图的）上（下）一半
top（bottom）row	上（下）一排
top（middle，bottom，right，left）panel（plot）	指图中某一小（分）图的位置
blackened（full，filled，solid）circle	实心圆
open circle	空心圆
line of circles	圆点组成的线
（solid，open）square	（实心、空心）方块
cross	十字符号
dashed line（chain dash）	小线段（虚线）
dash-dot-dash line	点划线（—·—）
chain dot	点线（……）
dotted-dashed line	点划线
（solid，broken）line	（实、虚）线
heavy（thick）solid line	粗实线
thin（light）broken line	细虚线
（straight，wavy）line	（直、波状）线
（smooth，dotted）curve	（平滑、点）曲线
（shaded，clear）area	（阴影、空白）区
（dotted，hatched，cross-hatched）area	（布点、网状、阴影线）区
（dark，light）shaded area	（深、线）阴影区

② 表

与表有关的词汇有：table，form，list 等。表的编号、标题的位置以及对表的来源的说明等与图类似。注意，英语的表格一般只列横线，尽量少列竖线，几乎没有斜线。当一页不能容纳一张表时，则在当页下注明 to be continued 并在下页上注明 continued。另外，对表中项目的注释可放在表中，也可放在表外。

在解释表格内容时，会论及 row（行）、column（列）等，例如：

two rows from top	前两行

the middle row	中间 1 行
the third column from right	右数第 3 列
the second row from bottom	倒数第 2 行

③ 公式

公式或方程在科技论文中比比皆是。如同图表编号，公式的编号可按顺序依次进行，或按章节分开进行，其位置一般在公式的右侧靠边。下面给出若干描述公式推导的例子和相关词汇。

［例 44］ 假设……，从……中得到结论。

Assuming that...，the solution takes the from...

其中，可用 by setting（putting，letting）... 替代 assuming that...，与 take 相近的词有：result in，yield，give，get，have，arrive at，find，obtain，produce，follow 等。

［例 45］ 通过方程（2）类推，方程可以重新写成……形式。

By analogy to Eq.（2）the equation can be rewritten in the form of...

其中 by analogy to（by analogy with，on the analogy of ）表示“根据……类推”。in the form（of the form，in... form）表示“以……形式”，如

in linear form	以线性形式
in equation form	以方程形式
in finite-difference form	以有限差分形式
in symbolic form	以符号形式
in nondimensional form	以无量纲形式
in integral form	以积分形式
in vectorial form	以矢量形式

［例 46］ 将方程（3）代入方程（5），可以得到……

Substituting Eq.（3）into Eq.（5）it follows that...

［例 47］ 合并方程（1）和方程（2），使我们写出应力的表达式 $\sigma=xyz$。

Combining Eq.（1）and Eq.（2）allows us to write the expression for stress $\sigma=xyz$.

推导方程时常用到以下短语：

(by) substituting... into...

(by) inserting...

(by) eliminating... in...

(by) combining... and...

(by) introducing... into...

(by) multiplying（dividing）... by...

subtracting（adding）...

(by) solving

(by) neglecting（ignoring，dropping）...

［例 48］ 式中 x 代表……，y 代表……，z 代表……

Where（in which）x refers to...，y is... and $z=$...

若需要对公式中的数学符号进行解释和定义，应在公式下一行起头处，用 where 或 in which（式中）引出。

［例 49］ 在初始条件下，方程（5）和方程（6）被确定……

Eq.（5）and Eq.（6）were performed on initial conditions that...

常用的相关词汇有：perform，proceed，derive，simplification，approximation，(re)-arrangement，algebra，positive，negative，condition，assumption 等。

(9) Quantity and Unit 数量与单位

① 对数量的描述

［例 50］ 已经做了许多试验，还需要进行许多观察。

Scores of experiments have been done and a lot of observations needed to be processed.

描述数量、次数等的短语有：

a lot of, lots of, a great many	许多
a large quantity of, a great deal of	许多
a negligible amount of	很少一点
an insufficient quantity of	量不大的
a wide variety of	各种各样的
a mass of, a volume of, a world of	大量的
a series of, a train of	一系列的

［例 51］ 桥上行人产生的活荷载大约 3.5kN/m^2。

The live load caused by passengers on bridges is about 3.5kN/m^2.

描述数量大小、近似、范围等的短评和词汇有：

a bout, around, some, roughly, approximately	大约
in the vicinity of, a matter of, in the neighborhood of	大约
of (in, on) the order of	大约
order of magnitude	量级
range from... to..., in the range of	……在……范围内变化
up to...	最大（高、多……）达……
down to...	最小（低、少……）达……
as high (low, many, few) as	……高（低、多、少……）达……
in excess of, over, above, more (greater, higher) than	超过，……以上
below, under, less (fewer) than	低于，……以下
increase (differ, decrease, change) by...	增加（相差、减小、变化）……

② 度量衡和单位换算

在论文写作中，常常会用到度量衡和单位换算。一般在论文中采用以下两种方式处理：一是用两种度量标注数量，如 2.5kip/ft（36.5kN/m）；二是采用一种单位制，但在论文中或附录中列出所用到的单位换算。土木工程常用的度量衡和单位换算见附录 3。

③ 单位的表达

在英语中，“以……为单位”为 in unit of... 或 in...，by...。例如：

an angle in radians	以弧度为单位的角
weight in tons	重量以吨计
vehicle speed in m/s	车速的单位为 m/s
by weight (volume, area)	以重量（体积、面积）计

当以数量词为单位时，采用“in＋基数词的复数”形式，如：

in hundreds	以百个为单位
in thousands	以千个为单位
in dozens	以打为单位
in millions of US dollars	以百万美元为单位

(10) Normal Words and Phrases 常用词汇

调查与研究：investigate, inquire, explore, examine, look into, inspect, study, consider, search, seek, seek out, analysis 等。

设计与准备：design，scheme，project，plan，propose，arrange，dispose，organize 等。

实验与试验：experiment，test，trial，try out，measure，record，equipment 等。

处理与操作：examine，deal with，handle，treat，process，sort out，operate，conduct，activate，control，manage，function 等。

举例和例外：example，instance，case，illustration，exception，exclusion 等。

极值和均值：maximum，upper（上限），minimum，lower（下限），average 等。

准确和精确：accurate（准确），precise（精确），correct（正确），exact 等。

Part Ⅱ Collection of English Literatures about Engineerings

Unit 1 Careers in Civil Engineering

Engineering is a profession, which means that an engineers must have a specialized university education. ❶ Many government jurisdictions also have licencing procedures which require engineering graduates to pass an examination, similar to the bar examination for a lawyer, before they can actively start on their careers.

In the university, mathematics, physics, and chemistry are heavily emphasized throughout the engineering curriculum, but particularly in the first two or three years. Mathematics is very important in all branches of engineering, so it is greatly stressed. Today, mathematics includes courses in statistics, which deals with gathering, classifying, and using numerical data, or pieces of information. An important aspect of statistical mathematics is probability, which deals with what may happen when there are different factors, or variables, that can change the results of a problem. Before the construction of a bridge is undertaken, for example, a statistical study is made of the amount of traffic the bridge will be expected to handle. ❷ In the design of the bridge, variables such as water pressure on the foundations, impact, the effects of different wind forces, and many other factors must be considered.

Because a great deal of calculation is involved in solving these problems, computer programming is now included in almost all engineering curricula. Computers, of course, can solve many problems involving calculations with greater speed and accuracy than a human being can. But computer are useless unless they are given clear and accurate instructions and information, in other words, a good program.

In spite of the heavy emphasis on technical subjects in the engineering curriculum, a current trend is to require students to take courses in the social sciences and the language arts. The relationship between engineering and society is getting closer; it is sufficient, therefore, to say again that the work performed by an engineer affects society in many different and important ways that he or she should be aware of. ❸ An engineer also needs a sufficient command of language to be able to prepare reports that are clear and, in many cases, persuasive. An engineer engaged in research will need to be able to write up his or her findings for scientific publications.

The last two years of an engineering program include subjects within the student's field of specialization. For the student who is preparing to become a civil engineer, these specialized courses may deal with such subjects as geodetic surveying, soil mechanics, or hydraulics.

Active recruiting for engineers often begins before the student's last year in the university. Many different corporations and government agencies have competed for the services of engineers in recent years. In the science-oriented society of today, people who have technical training are, of course, in demand. Young engineers may choose to go into environmental or sanitary engineering, for example, where environmental concerns have created many openings;[4] or they may choose construction firms that specialize in highway work; or they may prefer to work with one of the government agencies that deals with water resources. Indeed, the choice is large and varied.

When the young engineering has finally started actual practice, the theoretical knowledge acquired in the university must be applied. He or she will probably be assigned at the beginning to work with a team of engineers. Thus, on-the-job training can be acquitted that will demonstrate his or her ability to translate theory into practice to the supervisors.[5]

The civil engineer may work in research, design, construction supervision, maintenance, or even in sales or management. Each of these areas involves different duties, different emphases, and different uses of the engineer's knowledge and experience.

Research is one of the most important aspects of scientific and engineering practice. A researcher usually works as a member of a team with other scientists and engineers. He or she is often employed in a laboratory that is financed by government or industry. Areas of research connected with civil engineering include soil mechanics and soil stabilization techniques, and also the development and testing of new structural materials.

Civil engineering projects are almost always unique; that is, each has its own problems and design features. Therefore, careful study is given to each project even before design work begins. The study includes a survey both of topographical and subsoil features of the proposed site. It also includes a consideration of possible alternatives, such as a concrete gravity dam or an earth-fill embankment dam. The economic factors involved in each of the possible alternatives must also be weighed. Today, a study usually includes a consideration of the environmental impact of the project. Many engineers, usually working as a team that includes surveyors, specialists in soil mechanics, and experts in design and construction, are involved in making these feasibility studies.

Many civil engineers, among them the top people in the field, work in design. As we have seen, civil engineers work on many different kinds of structures, so it is normal practice for an engineer to specialize in just one kind. In designing buildings, engineers often work as consultants to architectural or construction firms. Dams, bridges, water supply systems, and other large projects ordinarily employ several engineers whose work is coordinated by a systems engineer who is in charge of the entire project. In many cases, engineers from other disciplines are involved. In a dam project, for example, electrical and mechanical engineers work on the design of powerhouse and its equipment. In other cases, civil engineers are assigned to work on a project in another field; in the space program, for instance, civil engineers were necessary in the design and construction of such structures as launching pads and rocket storage facilities.

Construction is a complicated process on almost all engineering projects. It involves scheduling the work and utilizing the equipment and the materials so that costs are kept as low as possible. Safety factors must also be taken into account, since construction can be very dangerous. Many civil engineers therefore specialize in the construction phase.

Much of the work of civil engineers is carried on outdoors, often in rugged and difficult terrain or under dangerous conditions. Surveying is an outdoors occupation, for example, and dams are often built in wild river valleys or gorges. Bridges, tunnel, and skyscrapers under construction can also be dangerous places to work. In addition, the work must also progress under all kinds of weather conditions. The prospective civil engineer should be aware of the physical demands that will be made on him or her.

Words and Phrases

1	jurisdiction	*n.*	管辖权，权限
2	government jurisdiction		政府行政区
3	bar	*n.*	法庭，律师的职业
4	curriculum	*n.*	课程，学习计划
5	probability	*n.*	概率论，可能性
6	impact	*n.*	冲击力
7	geodetic	*n.*	大地测量学
8	hydraulics	*n.*	水利学
9	recruit	*v.*	招聘
10	orient	*n.*	定向，定位
11	science-orient		注重科学的
12	supervision	*n.*	管理，监控
13	construction	*n.*	施工，建设
14	topographic (al)	*a.*	地形学（的）
15	subsoil	*n.*	下（亚）层土，地基下层土
16	alternative	*n.*	比较方案
		a.	交替的，变更的，比较的
17	consultant	*n.*	顾问，咨询者
18	architectural	*a.*	建筑（学）的
19	rugged	*a.*	崎岖的，艰难的
20	terrain	*n.*	地域，地带，领域
21	gorge	*n.*	峡谷
22	engineering graduate		工科毕业生
23	civil engineer		土木工程
24	scientific publication		科学刊物
25	geodetic surveying		大地测量学
26	soil mechanics		土力学
27	on-the-job		在现场的，在职的
28	soil stabilization		土壤稳定
29	structural materials		建筑材料
30	earth-fill embankment dam		填土坝
31	feasibility study		可行性研究

32 launching pads 发射台
33 rocket storage facilities 火箭库
34 construction phase 施工阶段

Notes

❶ ...，which means that... 这是一个非限定性定语从句。关系代词代表前面整个主句 engineering is a profession。全句可分成两句：工程是一种专业，这就是说工程师必须受过专业大学教育。

❷ ... of the amount traffic the bridge will be... 句中介词短语 of the amount traffic 作定语用，修饰 a statistical study。因为它比较长，而谓语动词较短，故将其移到谓语之后。The bridge will be... 是个定语从句，修饰 traffic，此处省略了关系代词"which"。

❸ It is sufficient，therefore，to say... 句中的 it 是形式主语，真实主语是不定式 to say again that...，其中 that the work... ways 是 to say 的宾语从句，后面的 that he or she should be aware of 是定语从句，修饰 ways。

❹ ... where environmental concerns have created many openings. 这个定语从句修饰 environmental or sanitary engineering。

❺ that will demonstrate his or her ability... to the supervisors. 这是定语从句，用来修饰 on-the-job training，翻译时可将从句另外译成：这将使主管人了解他（她）的……的能力。

Exercises

1. Translate the following into Chinese

(1) Before the construction of a bridge is undertaken，for example，a statistical study is made of the amount of traffic the bridge will be expected to handle.

(2) The relationship between engineering and society is getting closer；it is sufficient，therefore，to say again that the work performed by an engineer affects society in many different and important ways that he or she should be aware of.

2. Translate the following into English

(1) 需要强调数学、力学和计算机技术在土木工程应用中的重要性。

(2) 总工程师负责项目的构思与规划。

(3) 在许多情况下，可能会指派土木工程师参与其他工程项目的工作。

Unit 2　Modern Buildings and Structural Materials

Many great buildings built in earlier ages are still in existence and in use. Among them are the Pantheon and Colosseum in Rome, Hagia Sophia in Istanbul; the Gothic churches of France and England, and the Renaissance cathedrals, with their great domes, like the Duomo in Florence and St. Peter's in Rome. ❶ They are massive structures with thick stone walls that counteract the thrust of their great weight. Thrust is the pressure exerted by each part of a structure on its other parts.

These great buildings were not the product of knowledge of mathematics and physics. They were constructed instead on the basis of experience and observation, often as the result of trial and error. One of the reasons they have survived is because of the great strength that was built into them—strength greater than necessary in most cases. ❷ But the engineers of earlier times also had their failure. In Rome, for example, most of the people lived in insula, great tenement blocks that were often ten stories high. Many of them were poorly constructed and sometimes collapsed with considerable loss of life.

Today, however, the engineer has the advantage not only of empirical information, but also of scientific data that permit him to make careful calculations in advance. When a modern engineer plans a structure, he takes into account the total weight of all its component materials. This is known as the dead load, which is the weight of the structure itself. He must also consider the live load, the weight of all the people, cars, furniture, machines, and so on that the structure will support when it is in use. In structures such as bridges that will handle fast automobile traffic, he must consider the impact, the force at which the live load will be exerted on the structure. He must also determine the safety factor, that is, an additional capability to make the structure stronger than the combination of the three other factors.

The modern engineer must also understand the different stresses to which the materials in a structure are subject. These include the opposite forces of compression and tension. In compression the material is pressed or pushed together; in tension the material is pulled apart or stretched, like a rubber band. In addition to tension and compression, another force is at work, namely shear, which we defined as the tendency of a material to fracture along the lines of stress. The shear might occur in a vertical plane, but it also might run along the horizontal axis of the beam, the neutral plane, where there is neither tension nor compression.

Altogether, three forces can act on a structure: vertical—those that act up or down; horizontal—those that act sideways; and those that act upon it with a rotating or turning motion. Forces that act at an angle are a combination of horizontal and vertical forces. Since the structures designed by civil engineers are intended to be stationary or stable, these forces must be kept in balance. The vertical forces, for example, must be equal to each other. If a beam supports a load above, the beam itself must have sufficient strength to counterbalance that weight. The horizontal forces must also equal each other so that there is not too much thrust either to the right or to the left. And forces that might pull the structure around must he countered with forces that pull in the opposite direction.

One of the most spectacular engineering failures of modern time, the collapse of the Tacoma Narrows Bridge in 1940, was the result of not considering the last of these factors carefully enough. When strong gusts of wind, up to sixty-five kilometers an hour, struck the bridge during a storm, they set up waves along the roadway of the bridge and also a lateral motion that caused the roadway to fall. Fortunately, engineers learn from mistakes, so it is now common practice to test scale models of bridges in wind runnels for aerodynamic resistance.

The principal construction materials of earlier tines were wood and masonry brick, stone, or tile, and similar materials. The courses or layers were bound together with mortar or bitumen, a tar-like substance, or some other binding agent. The Greeks and Romans sometimes used iron rods or clamps to strengthen their buildings. The columns of the Parthenon in Athens, for example, have holes drilled in them for iron bars that have now rusted away. ❸ The Romans also used a natural cement called pozzolana, made from volcanic ash, that became as hard as stone under water.

Both steel and cement, the two most important construction materials of modern times, were introduced in the nineteenth century. Steel, basically an alloy of iron and a small amount of carbon, had been made up to that tine by a laborious process that restricted it to such special uses as sword blades. After the invention of the Bessemer process❹ in 1856, steel was available in large quantities at low prices. The enormous advantage of steel is its tensile strength; that is, it does not lose its strength when it is under a calculated degree of tension, a force which, as we have seen, tends to pull apart many materials. ❺ New alloys have further increased the strength of steel and eliminated some of its problems, such as fatigue, which is a tendency for it to weaken as a result of continual changes in stress.

Modern cement, called Portland cement, was invented in 1824. It is a mixture of limestone and clay, which is heated and then ground into a powder. It is mixed at or near the construction site with sand, aggregate (small stones, crushed rock, or gravel), and water to make concrete. Different proportions of the ingredients produce concrete with different strength and weight. Concrete is very versatile; it can be poured, pumped, or even sprayed into all kinds of shapes. And whereas steel has great tensile strength, concrete has great strength under compression. Thus, the two substances complement each other.

They also complement each other in another way: they have almost the same rate of contraction and expansion. They therefore can work together in situations where both compression and tension are factors. Steel rods are embedded in concrete to make reinforced concrete in concrete beams or structures where tension will develop. Concrete and steel also form such a strong bond-the force that unites them-that the steel cannot slip within the concrete. Still another advantage is that steel does not rust in concrete. Acid corrodes steel, whereas concrete has an alkaline chemical reaction, the opposite if acid.

Prestressed concrete is an improved form of reinforcement. Steel rods are bent into the shapes to give them the necessary degree of tensile strength. They are then used to prestress concrete, usually by pretensioning or posttensioning method. Prestressed concrete has made it possible to develop buildings with unusual shapes, like some of the modern sports arenas, with large spaces unbroken by any obstructing supports. ❻ The uses for this relatively new structural method are constantly being developed.

The current tendency is to develop lighter materials. Aluminum, for example, weighs much less than steel but has many of the same properties. Aluminum beams have already been uses for bridge construction and for the framework of a few buildings.

Attempts are also being made to produce concrete with more strength and durability, and with a lighter weight. One system that helps cut concrete weight to some extent uses polymers, which are long chainlike compounds used in plastics, as part of the mixture. ❼

Words and Phrases

1	counteract	*v.*	抵抗
2	insula	*n.*	群房，公寓
3	thrust	*n.*, *v.*	推，推力，轴力
4	tenement	*n.*	出租的房子，经济公寓
5	concave	*a.*, *n.*	凹的，凹面
6	convex	*a.*, *n.*	凸的，凸面
7	shear	*n.*, *v.*	剪切，剪力
8	roadway	*n.*	车行道，路面
9	masonry	*n.*	圬工，砌筑
10	mortar	*n.*	砂浆，灰浆
11	bitumen	*n.*	沥青
12	tar-like	*a.*	焦油般的
13	clamp	*n.*, *v.*	夹子，夹钳；卡紧
14	cement	*n.*	水泥
15	aggregate	*n.*	骨料，集料
16	ingredient	*n.*	（混合物）成分，配料
17	versatile	*a.*	多用途的，多方面适应的
18	alkaline	*a.*	碱性（的）
19	arena	*n.*	表演场
20	polymer	*n.*	聚合物
21	fatigue	*n.*	疲劳
22	trail and error		反复试验，试错法，尝试法
23	dead load		恒载
24	live load		活载
25	volcanic ash		火山灰
26	safety factor		安全系数
27	neutral plane		中性面
28	rotating or turning moment		旋转力矩，扭转力矩

29	wind tunnel（test）	风洞（试验）
30	tensile strength	抗拉强度
31	binding agent	黏结料，结合料，黏合剂
32	Portland cement	波特兰水泥，硅酸盐水泥
33	construction site	施工现场
34	reinforced concrete	钢筋混凝土
35	prestressed concrete	预应力混凝土
36	pretensioning（posttensing）concrete	先（后）张法

Notes

❶ Pantheon 潘提翁神庙（公元前 124～公元前 120 年），位于意大利罗马；Colosseum 罗马大斗兽场（78～80 年）；Hagia Sophia 圣索菲亚教堂（533～537 年），位于土耳其伊斯坦布尔；Duomo（意大利语），意为 cathedral；St. Peter's 指罗马圣彼得大教堂（1506～1626 年），当时是在拉斐尔和米开朗琪罗等伟大艺术家们的亲自指导下建立起来的，是文艺复兴建筑中最完美的代表作。

❷ 比较级＋than（it is）necessary 超过了所需要的……此处省略了。Strength greater than necessary in most cases 是破折号前面 great strength 的同位语。

❸ Parthenon 帕提侬神庙，指古希腊雅典城邦的保护神雅典娜·帕提侬的神庙，是古希腊全盛时期建筑与雕刻的主要代表作。

❹ Bessemer process 贝色麦法，又称酸性底吹转炉炼钢法，由英国冶金学家 Henry Bessemer 在 1856 年首创。这是一种不需外热的、可大量生产的炼钢方法。

❺ a force which，as we have seen，tends to... 这是 a calculated degree of tension（特定程度的拉力）的同位语，as we have seen 是定语从句中的插入语。译为：就像我们已经知道的那种能把多种材料拉断的力。

❻ with large spaces unbroken by... 这个介词短语作状语用。译为：它们的大空间没有任何挡住视线的支撑物。

❼ One system that helps cut concrete weight to some extent uses polymers，which are long chainlike compounds used in plastics，as part of the mixture.

有一种用聚合物（塑料中用的长链化合物）作为部分配料的方法，这种方法有助于使混凝土的重量降到一定程度。

Exercises

1. Translate the following into Chinese

(1) In addition to tension and compression，another force is at work，namely shear，which we defined as the tendency of a material to fracture along the lines of stress.

(2) The enormous advantage of steel is its tensile strength；that is，it does not lose its strength when it is under a calculated degree of tension，a force which，as we have seen，tends to pull apart many materials.

2. Translate the following into English

(1) 目前，正在试图（attempt）生产出强度更高、耐久性更好且重量更轻的混凝土。

(2) 一般，材料承受（subject）拉力，或压力，或剪力，或这些力的组合作用。

Unit 3 Building Types and Design

A building is closely bound up with people, for it provides people with the necessary space to work and live in.

As classified by their use, buildings are mainly of two types: industrial buildings and civil buildings. Industrial buildings are used by various factories or industrial production while civil buildings are those that are used by people for dwelling, employment, education and other social activities.

Industrial buildings are factory buildings that are available for processing and manufacturing of various kinds, in such fields as the mining industry, the metallurgical industry, machine building, the chemical industry and the textile industry. Factory buildings can be classified into two types: single-story ones and multi-story ones. The construction of industrial buildings is the same as that of civil buildings. However, industrial and civil buildings differ in the materials used and in the way they are used.

Civil buildings are divided into two broad categories: residential buildings and public buildings. Residential buildings should suit family life. Each flat should consist of at least three necessary rooms: a living room, a kitchen and a toilet. Public buildings can be used in polities, cultural activities, administration work and other services, such as schools, office buildings, child-care centers, parks, hospitals, shops, stations, theatres, gymnasiums, hotels, exhibition halls, bath pools, and so on. All of them have different functions, which in turn require different design types as well.

Housing is the living quarters for human beings. The basic function of housing is to provide shelter from the elements,[1] but people today require much more than this of their housing. A family moving into a new neighborhood will want to know if the available housing meets its standards of safety, health, and comfort. A family will also ask how near the housing is to grain shops, food markets, schools, stores, the library, a movie theater, and the community center.

In the mid-1960's a most important value in housing was sufficient space both inside and out. A majority of families preferred single-family homes on about half an acre of land, which would provide space for spare-time activities. In highly industrialized countries, many families preferred to live as far out as possible from the center of a metropolitan area, even if the wage earners had to travel some distance to their work. Quite a large number of families preferred country housing to suburban housing because their chief aim was to get far away from noise, crowding, and confusion. The accessibility of pubic transportation had ceased to be a decisive factor in housing because most workers drove their cars to work. People were chiefly interested in the arrangement and size of rooms and the number of bedrooms.

Before any of the building can begin, plans have to be drawn to show what the building will be like, the exact place in which it is to go and how everything is to be done.

An important point in building design is the layout of rooms, which should provide the greatest possible convenience in relation to the purposes for which they are intended. In a dwelling house, the layout may be considered under three categories: "day" "night" and "services". Attention must be paid to the provision of easy communication between these areas. The "day" rooms generally include a dining-room, sitting-room and kitchen, but other rooms, such as a study, may be added, and there may be a hall. The living-room, which is generally the largest, often serves as a dining-room, too, or the kitchen may have a dining alcove. The "night" rooms consist of the bedrooms. The "services" comprise the kitchen, bathrooms, larder, and water-closets. The kitchen and larder connect the services with the day rooms.

It is also essential to consider the question of outlook from the various rooms, and those most in use should preferably face south as much as possible. It is, however, often very difficult to meet the optimum requirements, both on account of the surroundings and the location of the roads. In resolving these complex problems, it is also necessary to follow the local town-planning regulations which are concerned with pubic amenities, density of population, height of buildings, proportion of green space to dwellings, building lines, the general appearance of new properties in relation to the neighborhood, and so on.

There is little standardization in industrial buildings although such buildings still need to comply with local town-planning regulations. The modern trend is towards light, airy factory buildings with the offices, reception rooms, telephone exchange, etc., house in one low building overlooking the access road, the workshop, also light and airy, being less accessible to public view. ❷

Generally of reinforced concrete or metal construction, a factory can be given a "shed" type ridge roof, incorporating windows facing north so as to give evenly distributed natural lighting without sun-glare.

Words and Phrases

1	gymnasium	*n*.	体育馆，健身房
2	quarter	*n*.	住处
3	acre	*n*.	英亩（= 6.07 亩）
4	metropolitan	*a*.	大城市的
5	confusion	*n*.	混乱，杂乱
6	accessibility	*n*.	可达性，可接近性
7	layout	*n*.	计划，方案，布局，格式
8	study	*n*.	书房
9	alcove	*n*.	凹室，壁龛
10	larder	*n*.	食品室，储藏室
11	amenity	*n*.	舒适，适宜，愉快
12	shed	*n*.	小棚，小屋
13	ridge	*n*.	脊，岭
14	industrial (civil) buildings		工业（民用）建筑
15	spare-time activities		业余活动

16	building line	建筑红线
17	natural lighting	自然采光

Notes

❶ the elements 在此处表示自然力量，如风、雨等。

❷ The modern trend is... 一句可理解为由三个分句组成，各分句的主语分别为 the modern trend，house 和 workshop，后两个分句省略了谓语。

Exercises

1. Translate the following into Chinese

(1) Quite a large number of families preferred country housing to suburban housing because their chief aim was to get far away from noise，crowding，and confusion.

(2) The modern trend is towards light，airy factory buildings with the offices，reception rooms，telephone exchange，etc.，house in one low building overlooking the access road，the workshop，also light and airy，being less accessible to public view.

2. Translate the following into English

(1) 精心设计的建筑能为人们的工作和生活提供良好的空间环境。

(2) 即便远离工作地点，不少家庭仍宁愿选择位于郊区而不是城区的住房。

Unit 4　Introduction to Mechanics of Materials

Mechanics of materials is a branch of applied mechanics that deals with the behavior of solid bodies subjected to various types of loading. It is a field of study that is known by a variety of names, including "strength of materials" and "mechanics of deformable bodies". The solid bodies considered in this article include axially-loaded bars, shafts, beams, and columns, as well as structures that are assemblies of these components. Usually the objective of our analysis will be the determination of the stresses, strains, and deformations produced by the loads: if these quantities can be found for all values of load up to the failure load, ❶ then we will have obtained a complete picture of the mechanical behavior of the body.

Theoretical analyses and experimental results have equally important roles in the study of mechanics of materials. On many occasions we will make logical derivations to obtain formulas and equations for predicting mechanical behavior, but at the same time we must recognize that these formulas cannot be used in a realistic way unless certain properties of the material are known. These properties are available to us only after suitable experiments have been made in the laboratory. Also, many problems of importance in engineering cannot be handled efficiently by theoretical means, and experimental measurements become a practical necessity. The historical development of mechanics of materials is a fascinating blend of both theory and experiment, with experiments pointing the way to useful results in some instances and with theory doing so in others. ❷ Such famous men as Leonardo da Vinci (1452—1519) and Galileo Galilei (1564—1642) made experiments to determine the strength of wires, bars, and beams, although they did not develop any adequate theories (by today's standards) to explain their test results. By contrast, the famous mathematician Leonhard Euler (1707—1783) developed the mathematical theory of columns and calculated the critical load of a column in 1744, long before any experimental evidence existed to show the significance of his results. Thus, Euler's theoretical results remained unused for many years, although today they form the basis of column theory.

The concepts of stress and strain can be illustrated in an elementary way by considering the extension of a prismatic bar [see Fig. 1(a)]. A prismatic bar is one that has constant cross section throughout its length and a straight axis. In this illustration the bar is assumed to be loaded at its ends by axial forces P that produce a uniform stretching, or tension, of the bar. By making an artificial cut (section mm) through the bar at right angles to its axis, we can isolate part of the bar as a free body [Fig. 1(b)]. At the right-hand end the tensile force P is applied, and at the other end there are forces representing the action of the removed portion of the bar upon the part that remains. These forces will be continuously distributed over the cross section, analogous to the continuous distribution of hydrostatic pressure over a submerged surface. ❸ The intensity of force, that is, the force per unit area, is called the stress and is commonly denoted by the Greek letter σ. Assuming that the stress has a uniform

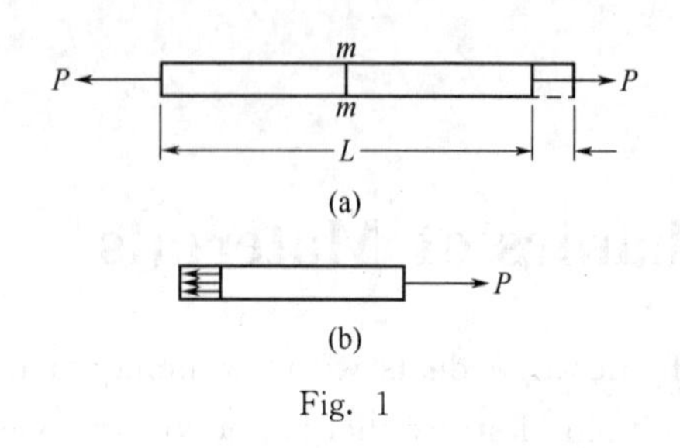

Fig. 1

distribution over the cross section [see Fig. 1(b)], we can readily see that its resultants is equal to the intensity to times the cross-sectional area A of the bar.

Furthermore, from the equilibrium of the body shown in Fig. 1(b), we can also see that this resultant must be equal in magnitude and opposite in direction to the force P. Hence, we obtain

$$\sigma = P/A \tag{1}$$

as the equation for the uniform stress in a prismatic bar. This equation shows that stress has units of force divided by area—for example, pounds per square inch (psi) or kips per square inch (ksi). When the bar is being stretched by the forces P, as shown in the figure, the resulting stress is a tensile stress, if the forces are reversed in direction, causing the bar to be compressed, they are called compressive stresses.

A necessary condition for Eq. (1) to be valid is that the stress σ must be uniform over the cross section of the bar. This condition will be realized if the axial force P acts through the centroid of the cross section, as can be demonstrated by static. ❹ When the load P does not act at the centroid, bending of the bar will result, and a more complicated analysis is necessary. Throughout this book, however, it is assumed that all axial forces are applied at the centroid of the cross section unless specifically stated to the contrary. Also, unless stated otherwise, ❺ it is generally assumed that the weight of the object itself is neglected, as was done when discussing the bar in Fig. 1.

The total elongation of a bar carrying an axial force will be denoted by the Greek letter δ [see Fig. 1(a)], and the elongation per unit length, or strain, is then determined by the equation

$$\varepsilon = \delta/L \tag{2}$$

where L is the total length of the bar. ❻ Note that the strain ε is a nondimensional quantity. It can be obtained accurately from Eq. (2) as long as the strain is uniform throughout the length of the bar. If the bar is in tension, the strain is a tensile strain, representing an elongation or stretching of the material; if the bar is in compression, the strain is a compressive strain, which means that adjacent cross sections of the bar move closer to one another.

Words and Phrases

1	blend	*n.*, *v.*	混合
2	elementary	*a.*	基本的，初步的
3	prismatic	*a.*	棱柱（形）的
4	analogous	*a.*	类似的，相似的
5	hydrostatic	*a.*	液体静力（学）的
6	intensity	*n.*	集度，强度
7	resultant	*n.*	合力
8	kip	*n.*	千磅（重量单位）
9	centroid	*n.*	形心，矩心
10	elongation	*n.*	伸长，延长

11	nondimensional	*a*.	无尺寸的，无单位的
12	analogous to		类似于，相似于
13	by contrast		对照之下

Notes

❶ ... all values of load up to the failure load. 句中 all values of load，所有荷载值；up to，达到；the failure load，断裂荷载。可译为：达到断裂荷载前的所有荷载值。

❷ with experiments pointing the way to useful results in some instances and with theory doing so in others. 句中两个 with 引出各自的独立结构，用 and 连接。doing 是代动词，指 pointing。第二个独立结构的完整形式是 with theory pointing the way to useful results in other instances.

❸ These forces will be continuously distributed over the cross section, analogous to the continuous distribution of hydrostatic pressure over a submerged surface. 句中 analogous to... 为形容词短语，说明句子主语 these forces。

❹ as can be demonstrated by... 句中 as 是关系代词，引出非限定性定语从句。As 代表整个主句所讲的内容，并在从句中作主语。译为：这一点可借静力学来证明。

❺ ... unless specifically stated to the contrary. Also, unless stated otherwise 句中 unless 引出省略的条件句，等于 unless it is stated... 译为：除非特殊说明不同情况。或除非另有说明。

❻ where L is the total length of the bar. 句中 where 是关系副词，引导非限定性定语从句，译为："其中"或"式中"。

Exercises

1. Translate the following into Chinese

(1) Mechanics of materials is a branch of applied mechanics that deals with the behavior of solid bodies subjected to various types of loading.

(2) This condition will be realized if the axial force P acts through the centroid of the cross section, as can be demonstrated by static. When the load P does not act at the centroid, bending of the bar will result, and a more complicated analysis is necessary.

2. Translate the following into English

(1) 通常，我们分析的目的是要确定由于荷载所产生的应力、应变和变形。

(2) 理论分析和试验结果在材料力学研究中具有同样重要的地位。

Unit 5 Loads

The loads for which a building must be designed may be classified into dead loads, vertical live loads, and lateral live and the weight of the include the weight of permanent equipment and the weight of the fixed components of the building, such as floors, beams, girders, roofs, columns, fixed walls and partition, and the like. All vertical loads other than dead loads may be included under vertical live loads. Lateral live loads are pressure. The probable effects of seismic disturbance are also lateral loads. The dynamic effects of crane movement may also be considered, in part, as lateral loads.

Dead Loads

The weight of the structural framing and all other building components fixed to and permanently supported by it are considered as dead loads. A reasonable estimate of the total weight of a building can usually be obtained from a preliminary sketch showing the structural and architectural layouts. This information is necessary for such steps in the analysis as the evaluation of the loads on the foundation. A correction to this actual calculation of the total weight shows a significant difference from the original assumptions.

For the design of the individual members, a more detailed dead load analysis usually is required. This involves a step-by-step procedure starting with the components directly supporting the live loads (floor and roof systems) and proceeding along the path of stresses (beams, girders, and columns) to the foundations. The dead loads on a member are determined only after the member itself and the portion of the structure it supports, have been designed. Thus, the actual dead load for each member should be checked and corrected; the design proceeds only after the necessary adjustments have been made.

Information of the weights of building materials is given in local building codes.

Vertical Live Loads

The loads on a building due to its occupancy, as well as snow loads on roof surfaces, are regarded as vertical live loads. Occupancy loads include personnel, furniture, machinery, stored materials, and other similar items. In buildings, live loads are often regarded as a uniformly distributed loading.

The amount of live load due to snow is dependent upon the location of the structure, the slope of the roof, and the orientation of the building with respect to wind direction. Snow load provisions are usually included in local building codes.

Several attempts have been made to standardize live load requirements due to occupancy, but building codes and specifications still have very diverse provisions on this subject.

The requirements are classified according to occupancy as follows:

(1) residential (including hotels)

(2) institutional (hospitals, sanitariums, jails)

(3) assembly (theatres, auditoriums, churches, schools)

(4) business (office-type building)

(5) mercantil (stores, shops, salesrooms)

(6) industrial (manufacturing, fabrication, assembly)

(7) storage (warehouses)

Load Combinations

In the design of roof trusses and single-story frames the loads carried by the structure include the dead load and the live load of snow and wind. The design usually is made on the basis of the stresses from the following combinations of loading:

Ⅰ dead load+snow

Ⅱ dead load+wind

Ⅲ dead load+wind+snow

Ⅳ dead load+wind+partial snow load

Partial snow load sometimes is considered due to drifting of snow. This may happen in building with uneven bay heights. Blown away but snow drift may accumulate on the lower roof level. It might even be that the snow load on the lower side of the building might be greater than normal.

Other load combinations may also be considered. A certain degree of engineering judgement is always required in selecting the proper load combinations.

In multi-story buildings, the stress is determined from a combination of:

Ⅰ dead load+live load

Ⅱ dead load+live load +wind or earthquake

Live loading is a random type of load and should be imposed on the structure such that maximum or critical moments and forces are induced on each individual member. From an analysis of the qualitative influence line for moment on the member of the multi-story frame, it can be shown that the combination of full dead load plus full live load may not always be the critical loading. The loading pattern shown if Fig. 23-1 usually called "checkerboard loading," may produce the critical conditions particularly for the maximum positive moment in the loaded spans.

Section [1. 5] and [2. 1] of the AISC specification allow an increase of 331/3 per cent in the allowable stressed, and a reduction of the load factor for plastic design from 1. 70 to 1. 30 when the stresses are induced by wind or earthquake forces. These provisions apply to load combinations Ⅱ, Ⅲ, and Ⅳ for roofs, and load combination Ⅱ for multi-story frames. The increase in allowable stress is also equivalent to considering three quarters of the value of the above load combinations. Thus, the choice of the critical stress should be based on a comparison of the values produced by the load combinations given in Table 1.

Table 1 load combinations

Building Component	Load Combination	Allowable Stress Design	Plastic Design
Roof	Ⅰ	$D+S$	$1.70(D+S)$
	Ⅱ	$3/4(D+W)$	$1.30(D+W)$
	Ⅲ	$3/4(D+W+S)$	$1.30(D+W+S)$
	Ⅳ	$3/4(D+W+\text{partial } S)$	$1.30(D+W+\text{partial } S)$
Floor	Ⅰ	$D+L$	$1.70(D+L)$
Multi-story Frame	Ⅰ Ⅱ	$D+L$ $3/4(D+L+W,\mathrm{E})$	$1.70(D+L)$ $1.30(D+L+W,\mathrm{E})$

Words and Phrases

1	lateral	*a.*	侧面的
2	permanent	*a.*	永久的，恒久的
3	partition	*n.*	隔墙
4	dynamic	*a.*	动力的
5	uniformly	*ad.*	均匀一致地
6	evaluation	*n.*	评估，评价
7	involve	*v.*	包括，包含
8	orientation	*n.*	方位，方向
9	mercantile	*a.*	商业的，贸易的

Exercises

1. Translate the following into Chinese

(1) A reasonable estimate of the total weight of a building can usually be obtained from a preliminary sketch showing the structural and architectural layouts.

(2) The amount of live load due to snow is dependent upon the location of the structure, the slope of the roof, and the orientation of the building with respect to wind direction.

2. Translate the following into English

(1) 在设计单个构件时，通常需要更为详细的恒载分析。

(2) 作用在结构上的活荷载常被看作均匀分布荷载。

Unit 6 Subsoils and Foundations

Loadings in buildings consist of the combined dead and imposed loads, which exert a downward pressure upon the soil on which the structure is founded and this in turn promotes a reactive force in the form of an upward pressure from the soil. The structure is in effect sandwiched between these opposite pressures and the design of the building must be able to resist the resultant stresses set up within the structural members and the general building fabric. The supporting subsoil must be able to develop sufficient reactive force to give stability to the structure to prevent failure due to unequal settlement and to prevent failure of the subsoil due to shear. ❶ To enable a designer to select, design and detail a suitable foundation he must have adequate data regarding the nature of the soil on which the structure will be founded and this is normally obtained from a planned soil investigation programmer.

Soil investigation. Soil investigation is specific in its requirements whereas site investigation is all embracing, taking into account such factors as topography, location of existing services, means of access and any local restrictions. Soil investigation is a means of obtaining data regarding the properties and characteristics of subsoils by providing samples for testing or providing a means of access for visual inspection. The actual data required and the amount of capital which can be reasonably expended on any soil investigation programmer will depend upon the type of structure proposed and how much previous knowledge the designer has of a particular region or site. ❷

The main methods of soil investigation can be enumerated as follows:

(1) Trial pits-small contracts where foundation depths are not likely to exceed 3m.

(2) Boreholes-medium to large contracts with foundations up to 30m deep.

Classification of soils. Soils may be classified by any of the following methods:

(1) Physical properties; (2) Geological origin; (3) Chemical composition; (4) Particle size.

It has been established that the physical properties of soils can be closely associated with their particle size both of which are of importance to the foundation engineer, architect or designer. All soils can be defined as being coarse-grained or fine-grained each resulting in different properties.

Coarse-grained soils: these would include sands and graves having a low proportion of voids, negligible cohesion when dry, high permeability and slight compressibility, which takes place almost immediately upon the application of load.

Fine-grained soils: these include the cohesive silts and clays having a high proportion of voids, high cohesion, very low permeability and high compressibility which takes place slowly over a long period of time.

There are of course soils, which can be classified in between the two extremes described above. BS 1377❸ deals with the methods of testing soils and divides particle sizes as follows:

Clay particles	less than 0. 002 mm
Silt particles	between 0. 002 and 0. 06 mm
Sand particles	between 0. 06 and 2 mm
Gravel particles	between 2 and 60 mm
Cobbles	between 60 and 200 mm

The silt, sand and gravel particles are also further subdivided into fine, medium and coarse with particle size lying between the extremes quoted above.

Shear strength of soils. The resistance which can be offered by a soil to the sliding of one portion over another or its shear strength is of importance to the designer since it can be used to calculate the bearing capacity of a soil and the pressure it can exert on such members as timbering in excavations. Resistance to shear in a soil under load depends mainly upon its particle composition. If a soil is granular in form, the frictional resistance between the particles increases with the load applied and consequently its shear strength also increases with the magnitude of the applied load. Conversely clay particles being small develop no frictional resistance and therefore its shear strength will remain constant whatever the magnitude of the applied load. Intermediate soils such as sandy clays normally give only a slight increase in shear strength as the load is applied.

Compressibility. Another important property of soils which must be ascertained before a final choice of foundation type and design can be made is compressibility, and two factors must be taken into account:

(1) Rate at which compression takes place.

(2) Total amount of compression when full load is applied.

When dealing with non-cohesive soils such as sands and gravels the rate of compression will keep pace with the construction of the building and therefore when the structure is complete there should be no further settlement if the soil remains in the same state. A soil is compressed when loaded by the expulsion of air and/or water from the voids and by the natural rearrangement of the particles. In cohesive soils the voids are very often completely saturated with water which in itself is nearly incompressible and therefore compression of the soil can only take place by the water moving out of the voids thus allowing settlement of the particles. Expulsion of water from the voids within cohesive soils can occur but only at a very slow rate due mainly to the resistance offered by the plate-like particles of the soil through which it must flow. This gradual compressive movement of a soil called consolidation. Uniform settlement will not cause under damage to a structure but uneven settlement can cause progressive structural damage.

Foundation types. There are many ways in which foundations can be classified but one of the most common methods is by form resulting in four basic types thus:

(1) Strip foundations-light loadings particularly in domestic buildings. Heavier loadings can sometimes be founded on a reinforced concrete strip foundation.

(2) Raft foundations-light loadings, average loadings on soils with low bearing capacities and structures having a basement storey.

(3) Pad or isolated foundations-common method of providing the foundation for columns of framed structures and for the supporting members of portal frames.

(4) Pile foundations-method for structures where the loads have to be transmitted to a point at some distance below the general ground level.

Words and Phrases

1	subsoil	*n.*	下层土，底土，天然地基
2	settlement	*n.*	沉陷，沉降
3	topography	*n.*	地形（势，貌）
4	borehole	*n.*	钻孔
5	gravel	*n.*	砾石
6	void	*n.*	孔隙，孔隙率
7	cohesion	*n.*	黏结力，内聚力
8	permeability	*n.*	渗透，渗透性
9	silt	*n.*	淤泥，残渣
10	cobble	*n.*	石子，圆石块
11	excavation	*n.*	挖掘
12	granular	*a.*	颗粒状的
13	saturate	*v.*	浸透，渗透，使充满
14	consolidation	*n.*	渗压，加强
15	soil investigation		土质勘测
16	trial pits		试验坑
17	geological origin		地质成因
18	particle size		粒径
19	coarse (fine)-grained soils		粗（细）粒土
20	shear strength		抗剪强度
21	domestic building		民用建筑
22	portal frame		门式框架

Notes

❶ The supporting subsoil must... to develop... 作 able 的补语，to give... 作 force 的定语，to prevent... 作全句的目的状语。

❷ The actual data... 句中，which 引出定语从句，修饰 the amount of capital；在 how much... 宾语从句中，其主、谓语 the designer has 插在 knowledge of 之间。

❸ BS 为 British Standard（英国标准）的缩写，后面的数字为规范编号。

Exercises

1. Translate the following into Chinese

(1) Soil investigation is a means of obtaining data regarding the properties and characteristics of subsoils by providing samples for testing or providing a means of access for visual inspection.

(2) In cohesive soils the voids are very often completely saturated with water which in itself is nearly incompressible and therefore compression of the soil can only take place by the water moving out of the voids thus allowing settlement of the particles.

2. Translate the following into English

(1) 不均匀沉降会导致结构产生更大的破坏。

(2) 土壤的物理性质与其本身的粒径有着密切的联系。

Unit 7 Philosophy of Structural Design

A structural engineering project can be divided into three phases: planning, design, and construction.

Structural design involves determining the most suitable proportions of a structure and dimensioning the structural elements and details of which it is composed. This is the most highly technical and mathematical phase of a structural engineering project, but it cannot and certainly should not be conducted without being fully coordinated with the planning and construction phases of the project. The successful designer is at all times fully conscious of the various considerations that were involved in the preliminary planning for the structure and, likewise, of the various problems that may later be encountered in its construction. ❶

Specially, the structural design of any structure first involves the establishment of the loading and other design conditions that must be resisted by the structure and therefore must be considered in its design. Then comes the analysis (or computation) of the internal gross forces (thrust, shears, bending moments, and twisting moments), stress intensities, strains, deflections, and reactions produced by the loads, temperature, shrinkage, creep, or other design conditions. ❷ Finally comes the proportioning and selection of materials of the members and connections so as to resist adequately the effects produced by the design conditions. The criteria used to judge whether particular proportions will result in the desired behavior reflect accumulated knowledge (theory, field and model tests, and practical experience), intuition, and judgment. For most common civil engineering structures such as bridges and buildings, the usual practice in the past has been to design on the basis of a comparison of allowable stress intensities with those produced by the service loadings and other design conditions. ❸ This traditional basis for design is called elastic design because the allowable stress intensities are chosen in accordance with the concept that the stress or strain corresponding to the yield point of the material should not be exceeded at the most highly stressed points of the structure. Of course, the selection of the allowable stresses may also be modified by a consideration of the possibility of failure due to fatigue, bucking, or brittle fracture or by consideration of the permissible deflections of the structure.

Depending on the type of structure and the conditions involved, the stress intensities computed in the analytical model of the actual structure for the assumed design conditions may or may not be in close agreement with the stress intensities produced in the actual structure by the actual conditions to which it is exposed. ❹ The degree of correspondence is not important, provided that the computed stress intensities can be interpreted in terms of previous experience. The selection of the service conditions and the allowable stress intensities provides a margin of safety against failure. The selection of the magnitude of this margin depends on the degree of uncertainty regarding loading, analysis, design, materials, and construction and on the consequences of failure. For example, if an allowable tensile stress of

20000 psi is selected for structural steel with a yield stress of 33000 psi, the margin of safety (or factor of safety) provided against tensile yielding is 33000/20000, or 1.65.

The allowable-stress approach has an important disadvantage in that it does not provide a uniform overload capacity for all parts and all types of structures. ❺ As a result, there is today a rapidly growing tendency to base the design on the ultimate strength and serviceability of the structure, with the older allowable-stress approach serving as an alternative basis for design. The newer approach currently goes under the name of strength design in reinforce-concrete-design literature and plastic design in steel-design literature. When proportioning is done on the strength basis, the anticipated service loading is first multiplied by a suitable load factor (greater than 1), the magnitude of which depends upon the uncertainty of the loading, the possibility of its changing during the life of the structure, and, for a combination of loadings, the likelihood, frequency, and duration of the particular combination. ❻ In this approach for reinforced-concrete design, the theoretical capacity of a structural element is reduced by a capacity-reduction factor to provide for small adverse variations in material strengths, workmanship, and dimensions. The structure is then proportioned so that, depending on the governing conditions, the increased load would (1) cause a fatigue or a buckling or a brittle-fracture failure or (2) just produce yielding at one internal section (or simultaneous yielding at several sections) or (3) cause elastic-plastic displacement of the structure or (4) cause the entire structure to be on the point of collapse.

Proponents of this latter approach argue that it results in a more realistic design with a more accurately provided margin of strength over the anticipated service conditions. These improvements result from the fact that nonelastic and nonlinear effects that become significant in the vicinity of ultimate behavior of the structure can be accounted for.

In recent decades, there has been a growing concern among many prominent engineers that not only is the term "factor of safety" improper and unrealistic, but worse still a structural design philosophy based on this concept leads in most cases to an unduly conservative and therefore uneconomical design, and in some cases to an unconservative design with too high a probability of failure. ❼ They argue that there is no such thing as certainty, either of failure or of safety of a structure but only a probability of failure or a probability of safety. They feel, therefore, that the variations of the probability of survival or the probability of serviceability of a structure estimated. ❽ It may not yet be practical to apply this approach to the design of each individual structure. However, it is believed to be practical to do so in framing design rules and regulations. It is highly desirable that building codes and specifications plainly state the factors and corresponding probabilities that they imply.

Words and Phrases

1	philosophy	*n.*	（基本）原理，哲学，宗旨
2	dimension	*n.*，*v.*	尺寸，尺度；定尺寸
3	conduct	*n.*，*v.*	行为，操行；引导，管理，传导
4	preliminary	*a.*	预备的，初步的
5	encounter	*v.*，*n.*	遭遇，遇到，相遇
6	gross	*a.*，*n.*	总的，显著的，总额

7	twist	*v.*, *n.*	扭转，编织
8	intensity	*n.*	强度，密（集）度
9	intuition	*n.*	直觉，直观
10	buckling	*n.*	压曲，弯折
11	margin	*n.*	空白，边缘，极限，富余
12	serviceability	*n.*	有用性，适用性
13	non-elastic	*a.*	非弹性的
14	nonlinear	*a.*	非线性的
15	vicinity	*n.*	附近，接近
16	unduly	*ad.*	过度地，不适当地
17	code	*n.*	规范，法规（则），（代）码
18	specification	*n.*	（常 *pl.*）规范，详述，规格，说明书
19	preliminary planning		初步规划
20	internal (gross) force		（总）内力
21	twisting moment		扭矩
22	stress intensity		应力强度
23	field (mode) test		现场（模型）试验
24	allowable stress		允许应力
25	yield point		屈服点
26	brittle fracture		脆裂
27	margin of safety (factor of safety)		安全系数
28	tensile yielding		抗拉屈服
29	allowable stress approach		允许应力法
30	ultimate strength		极限强度
31	capacity-reduction factor		承载能力折减系数
32	elastic-plastic displacement		弹塑性位移
33	point of collapse (collapse point)		破坏点

Notes

❶ The successful... 这句中 and 连接两个由 of 引导的介词短语，that 引导的定语从句分别修饰其前置词 considerations 和 problems。

❷ Then comes the analysis... 采用全部倒装句，then 状语前置，谓语为 come (follow)。

❸ For most common civil engineering... 这句中 those 指代前面的 stress intensities，以避免重复。过去分词短语 produced by... 作后置定语，修饰 those。

❹ Depending on the type... 这句中"Depending on..."引导一个条件状语，主干结构为"the stress intensities...may or may not be in close agreement with the stress intensities..."，"computed in... conditions"是过去分词短语作定语修饰前一个 stress intensities，"produced..."引导的过去分词短语作定语修饰后一个 stress intensities。

❺ The allowable-stress approach has... 这句中 that 引出一个宾语从句。

❻ When proportioning is done on the... 句中 which 指代前文中的 a suitable load factor。

❼ In recent decades，there has been a... 这句中 that 引导了一个由“not only... but (also)”连接的同位语从句作 a growing concern 的同位语，因 not only 提前，从句应用倒装句；but (also) 引导的从句中，worse still 为插入语，主干结构为“structural design leads to... and to...”，to 后引导两个介词短语作并列宾语。

❽ They feel，therefore，that the variations... 这句中 that 引导了由 and 连接的两个并列宾语从句“the variations of... manner”和“the probability... estimated”，句中 estimated 前省略了谓语动词 should be。

Exercises

1. Translate the following into Chinese

(1) This is the most highly technical and mathematical phase of a structural engineering project，but it cannot and certainly should not-be conducted without being fully coordinated with the planning and construction phases of the project.

(2) The allowable-stress approach has an important disadvantage in that it does not provide a uniform overload capacity for all parts and all types of structures.

(3) In recent decades，there has been a growing concern among many prominent engineers that not only is the term "factor of safety" improper and unrealistic，but worse still a structural design philosophy based on this concept leads in most cases to an unduly conservative and therefore uneconomical design，and in some cases to an unduly conservative design with too high a probability of failure.

2. Translate the following into English

(1) 内力包括轴力、剪力、弯矩、扭矩等。

(2) 在进行结构分析时，应该考虑（account for）各种荷载及其组合的作用。

(3) 目前，通常采用统计方法来研究结构所承受的荷载以及结构抗力。

Unit 8 Safety Concepts

Safety Tasks for Engineers

The gravitational force on a structure can be divided into dead loads and live loads. Dead loads can be calculated accurately because they rarely change with time and are usually fixed in one place. Live loads are always variable and movable, so no exact figures can be calculated for these forces. Structures must also resist other types of forces, such as wind or earthquakes, which are extremely variable. It is impossible to predict accurately the magnitude of all the forces that act on a structure during its life; we can only predict from past experience the probable magnitude and frequency of the loads.

Engineers never design a structure so that the applied loads exactly equal the strength of the structure. This condition is too dangerous because we can never know the exact value of either the applied loads or the strength of the structure. Therefore, a number called a "factor of safety" is used. The safety factor is loads on the structure. This factor may range from 1. 1 (where there is little uncertainty) to perhaps 5 or 10 (where there is great uncertainty).

Another task for engineers is construction supervising. When an architect receives a commission for a building, he meets the client and discusses his requirements. After visiting the site, the architect draws up preliminary plans and, together with a rough estimate of the cost, submits them to the client for his approval. If the client suggests changes, the architect incorporates them into the final design, which shows the exact dimension of every part of the building. At this stage, several building contractors are invited to bid for the job of constructing the building. When they submit their tenders or prices, the architect assists his client in selecting the best one and helps him to draw up a contract between the client and the contractor.

Work now starts on the building. As construction proceeds, the architect makes periodic inspections to make sure that the building is being constructed according to his plans and that the materials specified in the contract are being used. During the building period, the client pays the bills from the contractor. Subsequently, the contractor completes the building and the client occupies it. For six months after completion there is a period known as the "defects liability period" . During this period, the contractor must correct any defects that appear in the fabric of the building. Finally, when all the defects have been corrected, the client has to accept or reject it, he takes full possession of the building.

Safety in Design

The concept of safety in the context of current design is complex. The danger or risks inherent in structural design arise from the designer limited knowledge of the exact environment in which the structure may have to operate and also of the material properties and performance of the structure itself. In the context of building design and construction, safety is sought by using the best design and construction techniques available, the best materials and construction expertise and the best environmental data. With the design calculations, margins are ensured more specifically by the use of appropriate numerical factors. For the purpose of this discussion, it is accepted that all practical controls on material properties, workmanship, details, and so forth, are exactly as specified by the designer. Such an assumption is necessary to establish any analytical (or more properly, synthetic) process of design in which formal controls over safety are incorporated.

Consider the limit state or the Load and Resistance Factor Design (LRFD) approach to design. These methods appear to place equal importance on meeting serviceability requirements as on meeting those for ultimate limit states. It is arguable, however, that the principal objective of design is to produce serviceable structures and further, that completed structures complying with serviceability requirements are ipso-facto safe. The questions arise therefore: what function does the establishment of ultimate limit states perform vis-à-vis the security of structures? It can be said that the requirement of stability under extreme load states provides a specific security margin over the serviceability condition of the explicitly considered load cases. The evaluation of margin does not, within the conventional meaning of such a factor, result in a partial safety factor. Indeed, this part of the safety control mechanism is more akin to design requirements in that it affects primarily the nature of structural behavior.

The explicit use of numerical factors in the design process is to circumvent the differences between the design model and its assumptions and the real structure and its environment. ❶ Factors in this class reflect ignorance of the real world. Consider the ideal "perfectly safe" building. Such a building would be one for which the material and structural properties were known exactly and could be maintained throughout its life and also for which environmental, loading, and support conditions were known and controlled within the prescribed levels. No factors would be required in the design of this building.

Traditionally, designers have incorporated factors into their analyses by assuming pessimistic values of basic variables, by the use of conservative analytical models and quasi-global factors at appropriate stages of a design. In such an approach to design, it is only possible to decide qualitatively whether the method is successful or not by a study of the nature and rate of failure in the population of buildings constructed according to its principles. There is no practical way of evaluating the actual factor for any given load case.

The introduction of a partial safety factor system into the design process requires the possession of a significant body of data related to the basic variables to be considered in a design. This data should allow the assessment of ignorance of each variable independently. Each factor can only reflect the knowledge of one specified parameter. The many other factors known to affect design such as modeling errors, can not be evaluated individually allowance for them has to be made by some lumped or global design factor. This class of factor differs from those associated directly with particular variables in that determination of magnitude involves evaluation by some form of calibration in relation to existing design practice. ❷ Thus the introduction of partial factors of safety requires two following conditions to be satisfied: a large body of well-chosen data related to each basic variable and clear and unequivocal rules to quantify individual partial factors of safety.

For pragmatic reasons, where data for a particular variable is not available, realistic values of basic variables must be used whether chosen intuitively or by some analytical process.

The partial factor of safety systems incorporated into some modern codes or standards do not comply with the second condition. The use, for example of pessimistic load values defeats the object of a partial factor of safety approach.

The justification for changing the format of code must be that it becomes more self consistent and conforms more properly with the design process and that new data and methods can be introduced in the future without requiring modification of the factors; also ideally it should be easier to use.

Words and Phrases

1	expertise	*n.*	专门技能，专门知识
2	margin	*n.*	空白，边缘，极限，富余
3	workmanship	*n.*	工艺，技巧
4	serviceability	*n.*	适用（性），耐用（性）
5	arguable	*a.*	可争辩的，可论证的
6	comply	*vi*	应允，遵照（常跟 with）
7	ipso-facto		照那个事实，根据事实本身
8	vis-à-vis	*ad.*, *prep.*	相对；与……相比较
9	explicitly	*ad.*	明晰地，明确地
10	mechanism	*n.*	机理（制），技巧，手法
11	akin	*a.*	类似的，同样的（常跟 to）
12	circumvent	*vt.*	回避，绕过，围绕
13	quasi-global	*a.*	准（拟，半）总体的
14	analytical	*a.*	分析的，解析的
15	assessment	*n.*	估计，评价
16	allowance	*n.*	留量，容差，补助
17	lumped	*a.*	整块的
18	magnitude	*n.*	大小，量，量值
19	calibration	*n.*	标度，校准
20	unequivocal	*a.*	不含糊的，明确的
21	quantify	*v.*	确定数量，用数量表示
22	pragmatic	*a.*	重实效的，实际的
23	realistic	*a.*	现实（主义的）
24	intuitively	*ad.*	直觉上，直观上
25	justification	*n.*	正当理由，认为正当

Notes

❶ between the design model and... and the real structure and... 介词短语中 between 在此处指 the design model and its assumptions 和 the real structure and its environment 之

间的差别。

❷ This class of factor differs from those associated directly with particular variables in that determination of magnitude involves evaluation by some form of calibration in relation to existing design practice. in that 为介词宾语从句，表示“在……方面”，in relation to 表示与……相关。

Exercises

1. Translate the following into Chinese

(1) Such a building would be one for which the material and structural properties were known exactly and could be maintained throughout its life and also for which environmental, loading, and support conditions were known and controlled within the prescribed levels.

(2) In such an approach to design, it is only possible to decide qualitatively whether the method is successful or not by a study of the nature and rate of failure in the population of buildings constructed according to its principles.

2. Translate the following into English

(1) 在缺陷责任期内，承包商必须改正出现在建筑结构上的任何缺陷，直到所有缺陷被改正委托人接受，他才完全拥有这个建筑。

(2) 如模型错误等人们了解的影响设计的其他系数不能单独估计允许值，因为他们必须通过总体的综合的设计因素确定。

Unit 9 Design Criteria for Tall Building

The construction of tall buildings is the result of urbanization as seen in America since the late nineteenth century when the fist skyscrapers were built. The industrialized society attracts more people to the cities, requiring more space for offices as well as for habitation. Tall buildings, however, require two basic technical ingredients. First, economic method of building a tall building must be found and second, a reliable and economic method of transporting people vertically through the building must be available. Even though the Otis elevator in the late 19th century provided the logical answer to vertical transportation, the structure still remained a very significant deterrent to building very tall buildings. The pioneering Chicago School of Architecture refined the use of beam-column frames, but still required high "premium for height" for buildings taller than say 20 stories. As a result, buildings built up to 1920s were mostly below stories. A heroic step was taken in 1930 when the Empire State Building was built with more or less conventional method of construction. This was no example of any economic breakthrough. From the time of Depression in the 30's hardly any high rise construction was undertaken. After World War Ⅱ the socioeconomic needs in the U. S. opened the way for more taller office and apartment buildings. But, these buildings could no longer be designed in the same manner and proportion as the earlier buildings, simply because the concept of uncluttered column-free space, the concept of climate control and communication in the buildings have drastically changed. All these changes in attitudes brought the era of a new architecture. The new demand and challenge was to create total urban environments.

Within the last few years, research on building materials such as reinforced concrete and structural steel have made great strides and opened horizons for more efficient use of these materials. The structural engineers and architects also have met the challenge to find efficient and economical new structural systems for various ranges and heights of buildings going all the way to well over 100 stories. Consequently, the process of selecting a structural system for a tall building has become more complicated than it ever was. Structural systems can no longer be considered as an independent entity. It must now be a part of the overall systems approach used for the total design of a building.

The process of choosing the structural system for a tall building depends on many criteria which are not always structural. Without the understanding of all of these significant criteria, the structural engineer would feel frustrated in his efforts. The process of designing a building is multi-discipline, and since structure is an integral part of the total scheme, the successful adoption of an innovation in concept will have a much harder time of for acceptance if the engineer does not have an understanding of these criteria. The following is a brief discussion of these design criteria.

Environmental Planning Consideration. In any environment the addition of a tall building will certainly influence the operation of the traffic system, as well as the flow of people in the entire neighborhood. Therefore, a tall building project must be resolved in terms of pedestrian, auto and other traffic, and also resolve the overall need for space at the ground level. While an open plaza may be a viable solution under certain climatic and planning conditions, it may not be a reasonable solution where weather conditions are extreme either for winter or for summer. In such cases alternate to open plazas must be considered as a pan of the total solution, and the structural system must respond to it.

The Overall Proportions of the Tower. The relationship of the building to the surrounding environment and other existing buildings and plazas may dictate the proportions of the tower itself. It is, of course, obvious that a flat narrow tower will have a higher height-to-width ratio, which normally would mean more lateral sway. The increase in height-to-width ratio will generally mean increase in premium for height caused by additional material required to reduce lateral sway, as well as to increase resistance to overturning. Therefore, where slender buildings is an important requirement it is essential to find alternate structural solutions of higher efficiency even though such a system may increase fabrication and construction costs relative to more standard forms of construction.

Permissible Floor Area Ratio. The construction of buildings in the city area is generally controlled by zoning. Zoning, in most cases, unfortunately, is the product of political and economic consideration. Zoning would normally allow a maximum number of square feet that can be built at any given site. In dense urban centers this becomes a critical consideration, particularly because in terms of high land cost the more area that can be built on a given piece of land, the more the potential for economic return. Under the free enterprise system, therefore, there is every reason to believe that investors would like to build the fill allowable floor area ratio, which will then mean a taller building than otherwise considered consistent with the environment. Here again, the challenge to the architects and engineers is to provide viable alternatives using different structural systems from which a more rational and a more satisfying solution can be chosen.

Inner Space Criteria. Only a few years back, column spacing of 20 ft. was accepted as a structural limitation that could not be overcome. Newer structural systems have given the architect and developers the choice to create larger column free spaces in the office, as well as residential buildings. In the office buildings for instance, 35 ft. clear spacing is now normally considered the minimum and most developers would not mind if they can get a 60 ft. clear span between the core and the exterior walls. It is this kind of design consideration that has led to the development of a number of structural systems which do not require any column in the space between the core and the exterior walls. For apartments and hotel buildings the reverse is normally the case, that is, the maximum distance from the corridor to the outer wall does not normally exceed 30 ft., because every room is generally required to have an exterior exposure. This kind of planning requirement means that apartment and hotel buildings cannot be wider than between 70 to 90 ft.. Although wider apartment buildings have been built, the width of such buildings are still far less than what would be used for efficient office buildings. This type of requirement, of course, immediately means the possibility of a higher height-to-width ratio for residential buildings, and consequent selection of structural systems different than for office buildings.

Climatic Considerations. Climatic considerations sometimes play a strong part in choosing the structural system. For extremely cold winter climate the general tendency is to provide larger, clear windows, perhaps because there is a greater need for the occupant to look out and enjoy the nature as much as possible. On the other hand, in hot tropical climate it is more economical to have less glass and more solid masonry or concrete surface. This coincides with lesser need for the occupants to look outside. Structural systems which inherently use large exterior bearing walls are, therefore, quite often welcome in such climatic conditions.

Structural Material Considerations. The selection of a system depends strictly on the local relative economies between the various structural materials. It is because of this reason that while, in one area a concrete structural system may be economical, in another area of the country a structural steel system may become more economical. Sometimes, of course, a combination of both materials in the same building may result in the optimum design.

Foundation Considerations. Tall buildings require a stable foundation. Where rock is directly available near the ground level, the choice of structural system or material is not affected. However, in many areas where, for instance, floating foundation is the only way to support a building, the total weight of the structure becomes very significant because it controls the total depth of excavation. In such cases, light weight construction is the obvious structural choice. This could lead to either a steel structural framing, or an all lightweight concrete construction, or a composite system depending on the local material economies.

Time of Construction. Although time of construction of a project seems to be an obvious factor in choosing the type of construction, often it is overlooked as such, and estimates to compare structural systems are made without any consideration of the total time of construction. This may have little effect where interim financing during construction is not a critical item. In most investment projects, however, this cannot be overlooked. If the time of construction of a 1.5 million sq. ft. building takes 6 months more than another competitive structural system, it may very well mean an equivalent loss of $ 1.5 million in the total construction cost. In choosing structural systems, therefore, this criterion should be considered and only eliminated, if specifically requested by the owner.

Words and Phrases

1	urbanization	*n.*	都市化，都市集中化
2	deterrent	*a.*	制止的，威慑的
		n.	制止物，威慑因素
3	premium	*n.*	奖金，额外费用，高级
		a.	质量改进的，特级的
4	depression	*n.*	降低，不景气，萧条期
5	socioeconomic	*a.*	社会经济（学）的
6	clutter	*n.*	混乱，杂乱，干扰
		v.	弄乱，使混乱
7	pedestrian	*n.*	行人，步行者
8	plaza	*n.*	广场，集市场所，大空地
9	clear span		净跨
10	clear spacing		净距

11	corridor	*n.*	走廊，通道，过道
12	tropical	*n.*，*a.*	回归线（*pl.*）热带；热带的
13	excavation	*n.*	挖掘，挖方，发掘
14	interim	*a.*	间歇的，暂时的
15	zoning	*n.*	分区，区域化，分地带

Exercises

1. Translate the following into Chinese

(1) The process of designing a building is multi-discipline, and since structure is an integral part of the total scheme, the successful adoption of an innovation in concept will have a much harder time of for acceptance if the engineer does not have an understanding of these criteria.

(2) Where slender buildings is an important requirement it is essential to find alternate structural solutions of higher efficiency even though such a system may increase fabrication and construction costs relative to more standard forms of construction.

2. Translate the following into English

(1) 在近几年内，如钢筋混凝土和结构钢等建筑材料研究已经取得了很大进展，并开辟了更有效使用这些材料的前景。

(2) 尽管项目的施工工期似乎在选择结构类型中不是明显的因素，经常被忽视，但是没有对总体施工工期的考虑，就无法估算不同结构体系。

Unit 10 Durability at Concrete Structures

Although normally considered by engineers the most durable and soundest of materials, concrete must, under certain conditions, be listed as vulnerable due to various causes which result in cracking, corrosion of the steel or the chemical deterioration of the paste and the aggregates. In recent years various examples of unsatisfactory durability of concrete structures have been reported. Especially alarming is the rapidly growing number of concrete structures exhibiting signs of premature deterioration.

The yearly costs of keeping such structures in satisfactory operation have increased continuously. Recent investigations on national as well as on international level, e. g. within OECD,❶ reveal that these costs doubled during the eighties and will treble during the nineties.

The increasing number of concrete structures with unsatisfactory durability has apparently caught concrete profession off-guard. Concrete structures represent enormous investments by the Society, and the level of costs and inconveniences, which may be encountered if the problems of durability are not tackled promptly, may lead Society to an unjust global miscrediting of concrete as a prime building material, and hence question the credibility of the profession. ❷

The international concrete profession is, therefore, challenged by acute demands to develop and implement rational measures of solving the present twofold problems of durability, namely:

—Find measures to ensure a satisfactory remaining lifetime of existing structures threatened by premature deterioration.

—Incorporate in new structures the knowledge, experience and new research findings, in order to monitor the structural durability, thus ensuring the required service performance of future concrete structures.

All people involved in the planning, design and construction process should have the possibility of obtaining a minimum understanding of the possible deterioration processes and the decisive influencing parameters. Such a basic knowledge is the precondition for the ability to take the right decisions at the right time in order to ensure the required durability of concrete structures.

Protection of Reinforcement. Steel in concrete is protected against corrosion by passivation insured by the alkalinity (pH value >12.5). This passive layer impedes the dissolution of the steel. Thus corrosion of reinforcement is impossible, even if all other conditions are fulfilled (mainly oxygen and moisture).

Due to carbonation or by the action of chloride-irons the pH value may be reduced locally or over greater surface areas. When the pH drops below 9 at the reinforcement or if chloride content exceeds a critical value, the passive layer and the corrosion protection is lost and corrosion of reinforcement is possible.

Carbonation of Concrete. Concrete is a porous material and the carbon dioxide (CO_2) from the air can penetrate into the interior of the concrete. The lime fraction [$Ca(OH)_2$] of the hydrated cement is transformed in a carbonate ($CaCO_3$) and the pH value will drop below 9.

As already mentioned, the CO_2 penetrates from the surface to the interior of the concrete. Consequently, the carbonation starts from the concrete surface and penetrates slowly to the interior of the concrete. The speed-determining process is the diffusion of CO_2 into concrete and the rate of carbonation (increase of carbonation depth with time) follows a square-root-time law.

It can be observed that steel is best protected by a thick cover of impermeable concrete.

Diffusion of Chloride into Concrete. Besides CO_2, chloride ions (originating from sea water or deicing salt) may penetrate via the pores to the interior of the concrete. Chloride diffusion, however, is a diffusion process taking place in totally or partly water-filled pores.

The limestone, (carbonated lime, $CaCO_3$) has a certain chemical and physical binding capacity for chloride ions, depending on the chloride concentration in the pore water. However, not all the chlorides can be bound. There will, on the contrary, always exist dissolution equilibrium between bound chlorides and free chloride ions. With regard to corrosion of the reinforcement only the free chloride ions in the pore water are of controlling influence. Therefore it is of importance that after carbonation bound chlorides are released against so that the chloride content in the pore water and, consequently, the corrosion risk caused by chlorides will increase considerably after carbonation of concrete. A critical value of chloride concentration in the pore water for the incipient danger of corrosion will be illustrated later.

As a result of the diffusion process, the chloride concentration will decrease from the surface to the interior of the concrete following a square-root-time law.

Due to wetting and drying of the concrete surface with chloride-containing water, an enrichment of chlorides in the surface layer is possible.

At the beginning of the wetting period, a relatively great amount of chloride-containing water will penetrate into the concrete by capillary suction. During the drying period, the water dries out and the chlorides remain in the concrete. This process may cause a high enrichment of chlorides in the drying and wetting zone of a concrete (example: splash zones of marine structures).

Therefore, the water penetration depth of a concrete and the porosity of the surface layer respectively, are of great importance, especially in relation to the thickness of the concrete cover.

Corrosion of Steel. The corrosion process can be separated into two single processes: cathodic and anodic. In the anodic process, the positive iron ions pass into solution and rust will be formed by the combined hydroxyl ions at the cathode and the free iron at the anode, if oxygen is present by diffusion through the concrete cover. Water is necessary to enable the electrolytic process to take place, with destruction of the passive layer.

As the steel surface anodically and cathodically acting areas may be situated either close together at locally separated places even over relatively great distances. Consequently, corrosion may occur in areas of the structure where the direct access of oxygen to the surface of the reinforcement is impeded, if the concrete is wet enough to render the electrolytic connection possible.

In the region of cracks carbonation as well as chlorides tend to penetrate faster into the concrete. The influence of crack width on the corrosion rate of the reinforcement is relatively small within the common range of "normal" crack widths (0.15~0.35mm).

All the processes influencing corrosion of reinforcement are more or less controlled by diffusion processes: carbonation (diffusion of O_2). The major parameters is thus the quality of the concrete covers (thickness, impermeability and crack patterns).

Cracking. The cracking of concrete structures can be characterized as either minor or serious in nature. Minor cracking is often classified as that which is superficial and has no particular adverse affect on the structure. This means that it does not affect the safety or functionality of the structure and little or no additional maintenance is required to keep it under control. Serious cracking, on the other hand, is that which can adversely affect the safety, functionality, appearance, maintenance and rehabilitation costs.

Cracking will occur whenever the tensile strain to which concrete is subjected exceeds the tensile strain capacity of the concrete. Strains can be generated by various mechanisms:

—Movements generated within the concrete (examples: drying shrinkage, alkali-aggregate reaction, freeze/thaw cycles etc.).

—Expansion of material embedded within the concrete (examples: corrosion of reinforcement).

—Externally imposed conditions (loadings, settlements).

Good engineering can significantly reduce the potential for serious cracking. Taking steps to avoid the problem in the first place is always preferable to later rehabilitation. In essence the potential for cracking of concrete can be substantially reduced by various means: (a) selecting of proper concreting materials, including cement, pozzolan, aggregates, admixtures, (b) designing to minimize differential restraints and re-entrant corners and other crack initiators, (c) designing joint locations to control volumetric changes and random cracking.

Concrete is subjected to many adverse reactants or parameters (physical or chemical) that can reduce its durability.

Physical: Freezing and thawing, abrasion and cavitation.

Chemical: Alkali-aggregate, corrosion, chloride sulfates, carbonation.

These factors create chemical changes, micro-cracking resulting often in macro cracks.

Freezing and thawing in non air-entrained concretes generate destructive hydraulic pressures and tensile strains at last 10 times larger than the ultimate tensile strain of concrete, 200 microns. Corrosion of the reinforcement causes tremendous expansion and cracks because corroded steel occupies 7 times the volume of un-corroded steel.

Sulfates in the soils or in the water surrounding the concrete members will react with the tricalcium aluminate of the cement and cause expansion. Sulfate-resistant cement with low calcium aluminate will restrict this type of action.

Reactive aggregates and high-alkali cements cause swelling, map cracking and popouts in Portland cement concrete. The literature points out four types of reactions that may result due to reactive aggregates: alkali-silica reaction, alkali-carbonate reaction, silicification of carbonate aggregate.

These reactions are baffling because natural aggregates in concrete are expected to be inert in the cement plus water phase. The alkali-aggregate reaction has been identified in the early 40's and since that time the presence of this reaction has been observed in most countries around the world causing great damages. ❸ As it is a world problem it will be discussed more extensively.

...

Three factors are necessary before the alkali-silica reaction can occur:

(1) A reactive aggregate must be used.

(2) Alkali in the cement must be high.

(3) the concrete must be partially or totally wet.

Obviously, to avoid the consequences of the alkali-silica reaction in concrete, three principles must be recognized:

(1) Do not use reactive aggregates.

(2) Do not use cement with high alkali content.

(3) If it becomes necessary to bring a reactive aggregate together with high-alkali cement, use an admixture (such as a pozzolan) that will mitigate the deleterious consequences of the reaction.

To avoid the use of a reactive aggregate, it becomes first necessary to know which aggregates are reactive. If it is known that an aggregate has a satisfactory performance characteristic, continue to use it. If there is doubt or if the performance characteristics of a specific aggregate are not known, certain ASTM tests should be made on the aggregate. These tests entail a petrography examination, a chemical test, and a proof test.

The second way of avoiding an alkali-silica reaction is to limit the alkali content in the cement to a value of 0.60 percent.

The third method of avoiding an alkali-silica reaction is to use a pozzolanic or other admixture to mitigate the effects of alkali-aggregate reactions.

Words and Phrases

1	vulnerable	*a*.	易受伤害的，有弱点的
2	deterioration	*n*.	恶化，退化
3	durability	*n*.	延性
4	passivation	*n*.	钝化作用
5	alkalinity	*n*.	（强）碱性，碱度
6	carbonation	*n*.	碳化
7	chloride	*n*.	氯化物，漂白粉
8	ion	*n*.	离子
9	porous	*a*.	多孔性的，能渗透的
10	lime	*n*.	石灰
11	carbonate	*n*.	碳酸盐
12	impermeable	*a*.	不能渗透的

13	superficial	*a*.	表面的，肤浅的
14	thaw	*v*.	融化，解冻
15	rehabilitation	*n*.	复原，更新
16	pozzolan	*n*.	火山灰
17	abrasion	*n*.	擦伤，磨损
18	sulfate	*n*.	硫酸盐（酯）
19	baffling	*a*.	阻碍……的，起阻碍作用的
20	inert	*a*.	惰性的，迟钝的
21	deleterious	*a*.	有害的
22	petrography	*a*.	岩石（学）的，岩相（学）的
23	existing structures		既有结构
24	deicing salt		除冰盐
25	concrete cover		混凝土保护层
26	alkali-aggregate reaction		碱-骨料反应
27	tricalcium aluminate		三钙铝酸盐
28	popouts		（混凝土表面）剥落

Notes

❶ Organization for Economic Cooperation and Development 经济合作和发展组织。

❷ Concrete structures represent... 一句由两个分句组成，后一个分句的主语为 the level of costs and inconveniences，并带有一个由 which 引导的非限定性定语从句。

❸ The alkali-aggregate reaction has been... 句中，causing 前省略了 the presence of this reaction has been。

Exercises

1. Translate the following into Chinese

(1) Concrete structures represent enormous investments by the society, and the level of costs and inconveniences, which may by encountered if the problems of durability are not tackled promptly, may lead Society to an unjust global miscrediting of concrete as a prime building material, and hence question the credibility of the profession.

(2) The alkali-aggregate reaction has been identified in the early 40's and since that time the presence of this reaction has been observed in most countries around the world causing great damages.

2. Translate the following into English

(1) 在一定条件下，混凝土的延性会受到开裂、钢筋锈蚀、化学侵蚀等不利因素的影响。

(2) 减少碱-骨料反应的方法之一：不使用活性大的骨料来拌制混凝土。

Unit 11 Prestressed Concrete

Concrete is strong in compression, but weak in tensions: its tensile strength varies from 8 to 14 percent of its compressive strength. Due to such a low tensile capacity, flexural cracks develop at early stages of loading. In order to reduce or prevent such cracks from developing, a concentric or eccentric force is imposed in the longitudinal direction of the structural element. This force prevents the cracks from developing by eliminating or considerably reducing the tensile stress at the critical midspan and support sections at service load, thereby❶ raising the bending, shear, and torsional capacities of the sections. The sections are then able to behave elastically, and almost the full capacity of the concrete in compression can be efficiently utilized across the entire depth of the concrete sections when all loads act on the structure.

Such an imposed longitudinal force is called a prestressed force, i. e., a compressive force that prestress the sections along the span of the structural element prior to the application of the transverse gravity dead and live loads or transient horizontal live loads. The type of prestressing force involved, together with its magnitude, are determined mainly on the basis of the type of system to be constructed and the span length and slenderness desired. Since the prestressing force is applied longitudinally along or parallel to axis of the member, the prestressing principle involved is commonly known as linear prestressing.

Circular prestressing, used in liquid containment tanks, pipes, and pressure reactor vessels, essentially follows the same basic principles, as does linear prestressing. The circumferential hoop, or "hugging" stress on the cylindrical or spherical structure, neutralizes the tensile stresses at the outer fibers of the curvilinear surface caused by the internal contained pressure.

From the preceding discussion, it is plain that permanent stresses in the prestressed structural member are created before the full dead and live loads are applied in order to eliminate or considerably reduce the net tensile stresses caused by these loads. With reinforced concrete, it is assumed that the tensile strength of the concrete is negligible and disregarded. This is because the tensile forces resulting from the bending moments are resisted by the bond created in the reinforcement process. Cracking and deflection are therefore essentially irrecoverable in reinforced concrete once the member has reached its limit state at service load.

The reinforcement in the reinforced concrete member does not exert any force of its own on the member, contrary to the action of prestressing steel. The steel required to produce the prestressing force in the prestressed member actively preload the member, permitting a relatively high controlled recovery of cracking and deflection. Once the flexural tensile strength of the concrete is exceeded, the prestressed member starts to act like a reinforced concrete element.

Prestressed member are shallower in depth than their reinforced concrete counterparts for the same span and loading conditions. In general, the depth of a prestressed concrete member is usually about 65 to 80 percent of the depth of the equivalent reinforced concrete member. Hence, the prestressed member requires less concrete, and about 20 to 35 percent of the amount of reinforcement. Unfortunately, this saving in material weight is balanced by the higher cost of the higher quality materials needed in prestressing. Also, regardless of the system used, prestressing operations themselves result in an added cost: formwork is more complex, since the geometry of prestressed sections is usually composed of flanged sections with thin webs.

In spite of these additional costs, if a large enough number of precast units are manufactured, the difference between at least the initial costs of prestressed and reinforced concrete systems is usually not very large. And the indirect long-term savings are quite substantial, because less maintenance is needed, a longer working life is possible due to better quality control of the concrete, and lighter foundations are achieved due to the smaller cumulative weight of the superstructure.

Once the beam span of reinforced concrete exceeds 70 to 90 feet (21.3 to 27.4m), the dead weight of the beam becomes excessive, resulting in heavier members and, consequently, greater long-term deflection and cracking. Thus, for larger spans, prestressed concrete becomes mandatory since arches are expensive to construct and do not perform as well due to the severe long-term shrinkage and creep they undergo. Very large spans such as segmental bridges or cable-stayed bridges can only be constructed through the use of prestressing.

Prestressed concrete is not a new concept, dating back to 1872, when P. H. Jackson, an engineer from California, patented a prestressing system that used a tie rod to construct beams or arches from individual blocks. After a long lapse of time during which little progress was made because of the unavailability of high-strength steel to overcome prestress losses, R. E. Dill of Alexandria, Nenraska, recognized the effect of the shrinkage and creep (transverse material flow) of concrete on the loss of prestress. He subsequently developed the idea that successive post-tensioning of unbonded rods would compensate for the time-dependent loss of stress in the rods due to the decrease in the length of the member because of creep and shrinkage. In the early 1920s, W. H. Hewett of Minneapolis developed the principles of circular prestressing. He hoop-stressed horizontal reinforcement around walls of concrete tanks through the use of turnbuckles to prevent cracking due to internal liquid pressure, thereby achieving watertightness. Thereafter, prestressing of tanks and pipes developed at an accelerated pace in the United States, with thousands of tanks for water, liquid, and gas storage built and much mileage of prestressed pressure pipe laid in the two to three decades that followed.

Linear prestressing had progressed a long way from the early days, in particular through the ingenuity of Eugene Freyssinet, who proposed in 1926-28 methods to overcome prestress losses through the use of high-strength and high-ductility steels. In 1940, he introduced the now well-known and well-accepted Freyssinet system. ❷

P. W. Abeles of England introduced and developed the concept of partial prestressing between the 1930s and 1960s. F. Leonhardt of Germany, V. Mikhailov of Russia, and

T. Y. Lin of the United States also contributed a great deal to the art and science of the design of prestressed concrete. Lin's load-balancing method deserves particular mention in this regard, as it considerably simplified the design process, particularly in continuous structures. These twentieth-century developments have led to the extensive use of prestressing throughout the world, and in the United States in particular.

Today, prestressed concrete is used in buildings, underground structures, TV towers, floating storage and offshore structures, power stations, nuclear reactor vessels, and numerous types of bridge systems including segmental and cable-stayed bridges, they demonstrate the versatility of the prestressing concept and its all-encompassing application. The success in the development and construction of all these structures has been due in no small measures to the advances in the technology of materials, particularly prestressing steel, and the accumulated knowledge in estimating the short-and long-term losses in the prestressing forces.

Words and Phrases

1	longitudinal	*a.*	纵向的
2	transverse	*a.*	横向的
3	transient	*a.*, *n.*	瞬间，瞬态
4	slenderness	*n.*	细长（度）
5	circumferential	*a.*	周围的，环形的，环绕的
6	hoop	*n.*, *v.*	箍筋；箍住
7	spherical	*a.*	球（形）的
8	irrecoverable	*a.*	不能恢复的
9	formwork	*n.*	模板，支模
10	flange	*n.*	（梁的）翼缘
11	web	*n.*	（梁的）腹板
12	superstructure	*n.*	上部结构
13	mandatory	*a.*	必须遵循的，命令的
14	turnbuckle	*n.*	松紧螺旋扣
15	ingenuity	*n.*	独创性，机灵
16	flexural crack		挠曲裂缝
17	critical section		临界截面
18	service (live, dead) load		使用（活，静）荷载
19	prestressing force		预应力
20	linear (circular) prestressing		线（环）形预应力
21	precast unit		预制构件
22	time-dependent		与时间有关的，时变
23	partial prestressing		部分预应力
24	cable-stayed bridge		斜拉桥

Notes

❶ thereby 与现在分词 raising... 连用一般表示结果。

❷ Freyssinet system 弗莱西奈式（预应力）体系，简称弗氏体系。

Exercises

1. Translate the following into Chinese

(1) This force prevents the cracks from developing by eliminating or considerably reducing the tensile stresses at the critical midspan and support sections at service load, thereby raising the bending shear, and torsional capacities of the sections.

(2) The success in the development and construction of all these structures has been due in no small measures to the advances in the technology of materials, particularly prestressing steel, and accumulated knowledge in estimating the short and long term losses in the prestressing forces.

2. Translate the following into English

(1) 混凝土的长期收缩和徐变对预应力损失产生很大影响。

(2) 必须采用高强度、高延性钢作为预加应力的材料。

Unit 12 Structure Steel

Introduction to Structural Steel

Advantages of Steel as a Structural Material. A person traveling in the United States might quite understandably decide that steel was the perfect structural material. He or she would see an endless number of steel bridges, buildings, towers, and other structures-comprising, in fact, a list too lengthy to enumerate. After seeing these numerous steel structures this traveler might be quite surprised to learn that steel was not economically made in the United States until late in the nineteenth century and the first wide-flange beams were not rolled until 1908.

The assumption of the perfection of this metal, perhaps the most versatile of structural materials, would appear to be even more reasonable when its great strength, light weight, ease of fabrication, and many other desirable properties are considered. These and other advantages of structural steel are discussed in detail in the following paragraphs.

High Strength. The high strength of steel per unit of weight means that structure weights will be small. This fact is of great importance for long-span bridges, tall buildings, and structures having poor foundation conditions.

Uniformity. The properties of steel do not change appreciably with time, as do those of a reinforced-concrete structure.

Elasticity. Steel behaves closer to design assumptions than most materials because it follows Hooke's law ill up to fairly high stresses. The moments of inertia of a steel structure can be definitely calculated, while the values obtained for a reinforced-concrete structure are rather indefinite.

Permanence. Steel frames that are properly maintained will last indefinitely. Research on some of the newer steels indicates that under certain conditions no painting maintenance whatsoever will be required.

Ductility. The property of a material by which it can withstand extensive deformation without failure under high tensile stresses is said to be its ductility. When a mild or low-carbon structural steel member is being tested in tension, a considerable reduction in cross section and a large amount of elongation will occur at the point of failure before the actual fracture occurs. A material that does not have this property is generally unacceptable and is probably hard and brittle and might break if subjected to a sudden shock.

In structural members under normal loads, high stress concentrations develop at various points. The ductile nature of the usual structural steels enables them to yield locally at those points, thus preventing premature failures. A further advantage of ductile structures is that when overloaded their large deflections give visible evidence of impending failure (sometimes jokingly referred to as "running time").

Fracture Toughness. Structural steels are tough, that is, they have both strength and ductility. A steel member loaded until it has large deformations will still be able to withstand large forces. This is a very important characteristic because it means that steel members can be subjected to large deformations during fabrication and erection without fracture-thus allowing them to be bent, hammered, sheared, and have holes punched in them without visible damage. The ability of a material to absorb energy in large amounts is called toughness.

Additions to Existing Structures. Steel structures are quite well suited to having additions made to them. New bays or even entire new wings can be added to existing steel frame buildings, and steel bridges may often be widened.

Miscellaneous. Several other important advantages of structural steel are: (a) ability to be fastened together by several simple connection devices including welds and bolts; (b) adaptation to prefabrication; (c) speed of erection; (d) ability to be rolled into a wide variety of sizes and shapes; (e) fatigue strength; (f) possible reuse after a structure is disassembled, and (g) scrap value, even though not reusable in its existing form. Steel is the ultimate recyclable material.

Disadvantages of Steel as a Structural Material. In general, steel has the following disadvantages. Maintenance Costs Most steels are susceptible to corrosion when freely exposed to air and water and must therefore be periodically painted. The use of weathering steels, in suitable design applications, tends to eliminate this cost.

Fireproofing Costs. Although structural members are incombustible, their strength is tremendously reduced at temperatures commonly reached in fires when the other materials in a building burn. Many disastrous fires have occurred in empty buildings where the only fuel for the fires was the buildings themselves. Furthermore, steel is an excellent heat conductor, nonfireproofed steel members may transmit enough heat from a burning section or compartment of a building to ignite materials with which they are in contact in adjoining sections of the building. As a result of these facts the steel frame of a building may have to be protected by materials with certain insulating characteristics, or the building may have to include a sprinkler system if it is to meet the building code requirements of the locality in question.

Susceptibility to Buckling. The longer and more slender the compression members, the greater the danger of buckling. As previously indicated, steel has a high strength per unit of weight, but when used for steel columns is not very economical sometimes because considerable material has to be used merely to stiffen the columns against buckling.

Fatigue. Another undesirable property of steel is that its strength may be reduced if it is subjected to a large number of stress reversals or even to a large number of variations of tensile stress. (We have fatigue problems only when tension is involved.) The present practice is to reduce the estimated strengths of such members if it is anticipated that they will have more than a prescribed number of cycles of stress variation.

Brittle Fracture. Under certain conditions steel may lose its ductility, and brittle fracture may occur at places of stress concentration. Fatigue-type loadings and very low temperatures aggravate the situation.

Stress-strain Relationships in Structural Steel

In order to understand the behavior of steel structures, it is absolutely essential for the designer to be familiar with the properties of steel. Stress-strain diagrams present valuable information necessary to understand how steel will behave in a given situation. Satisfactory steel design methods cannot be developed unless complete information is available concerning the stress-strain relationships of the material being used.

If a piece of ductile structural steel is subjected to a tensile force it will begin to elongate. If the tensile force is increased at a constant rate, the amount of elongation will increase constantly within certain limits. In other words, elongation will double when the stress goes from 6000 to 12000 psi (pounds per square inch). When the tensile stress reaches a value roughly equal to one-half of the ultimate strength of the steel, the elongation will begin to increase at a greater rate without a corresponding increase in the stress.

The largest stress for which Hooke's law applies or the highest point on the straight-line portion of the stress-strain diagram is the proportional limit. The largest stress that a material can withstand without being permanently deformed is called the elastic limit. This value is seldom actually measured and for most engineering materials including structural steel is synonymous with the proportional limit. For this reason the term proportional elastic limit is sometimes used.

The stress at which there is a decided increase in the elongation or strain without a corresponding increase in stress is said to be the yield stress. It is the first point on the stress strain diagram where a tangent to the curve is horizontal. The yield stress is probably the most important property of steel to the designer, as so many design procedures are based on this value. Beyond the yield stress there is a range in which a considerable increase in strain occurs without increase in stress. The strain that occurs before the yield stress is referred to as the elastic strain; the strain that occurs after the yield stress, with no increase in stress, is referred to as the plastic strain. Plastic strains are usually from 10 to 15 times the elastic strains.

Yielding of steel without stress increase may be thought to be a severe disadvantage when in actuality it is a very useful characteristic. It has often performed the wonderful service of preventing failure due to omissions or mistakes on the designer's part. Should the stress at one point in a ductile steel structure reach the yield point, that part of the structure will yield locally without stress increase, thus preventing premature failure. This ductility allows the stresses in a steel structure to be readjusted. Another way of describing this phenomenon is to say that very high stresses caused by fabrication, erection, or loading will tend to equalize themselves. It might also be said that a steel structure has a reserve of plastic strain that enables it to resist overloads and sudden shocks. If it did not have this ability, it might suddenly fracture, like glass or other vitreous substances. Following the plastic strain there is a range in which additional stress is necessary to produce additional strain.

This is called strain-hardening. This portion of the diagram is not too important to today's designer because the strains are so large. A familiar stress-strain diagram for mild or low-carbon structural steel is shown in Fig. 1. Only the initial part of the curve is shown here because of the great deformation which occurs before failure. At failure in the mild steels the total strains are from 150 to 200 times the elastic strains. The curve will actually continue up to its maximum stress value and then "tail off" before failure. A sharp reduction in the cross section of the member takes place (called "necking") followed by failure.

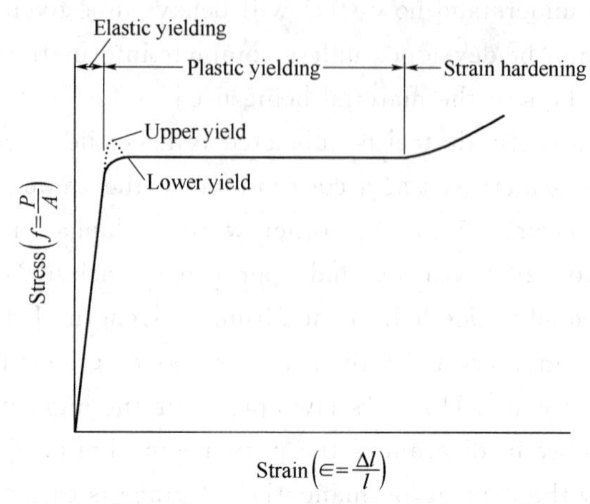

Fig. 1 Typical stress-strain diagram for a mild or low-carbon structural steel at room temperature

The stress-strain curve of Fig. 1 is typical of the usual ductile structural steel and is assumed to be the same for members in tension or compression. (The compression members must be stocky because slender compression members subjected to compression loads tend to bend laterally, and their properties are greatly affected by the bending moments so produced.) The shape of the diagram varies with the speed of loading, the type of steel, and the temperature. One such variation is shown in the figure by the dotted line marked upper yield. This shape stress-strain curve is the result when a mild steel has the load applied rapidly, while the lower yield is the case for slow loading.

You should note that the stress-strain diagram of Fig. 1 was prepared for mild steel at room temperature. Steels (particularly those with rather high carbon contents) may actually increase a little in strength as they are heated to a temperature of about 7000℉ As temperatures are raised into the 8000℉ to 10000℉ range, steel strengths are drastically reduced, and at 12000℉ they have little strength left.

Typical ratios of yield stresses at high temperatures to yield stresses at room temperatures are approximately 0.77 at 800℉, 0.63 at 1000℉, and 0.37 at 12000℉. Temperatures in these ranges can be easily reached in steel members during fires, in localized areas of members when welding is being performed, in members in foundries over open flame, and so on.

When steel sections are cooled below 320℉ their strengths will increase a little, but they will have substantial reductions in ductility and toughness.

A very important property of a structure which has not been stressed beyond its yield point is that it will return to its original length when the loads are removed. Should it be stressed beyond this point it will return only part of the way to its original position. This knowledge leads to the possibility of testing an existing structure by loading and unloading. If after the loads are removed the structure will not resume its original dimensions, it has been stressed beyond its yield point.

Steel is an alloy consisting almost entirely of iron (usually over 98 percent). It also contains small quantities of carbon, silicon, manganese, sulfur, phosphorus, and other elements. Carbon is the material that has the greatest effect on the properties of steel. The hardness and strength increase as the carbon percentage is increased, but unfortunately the resulting steel is more brittle and its weldability is adversely affected. A smaller amount of carbon will make the steel softer and more ductile, but also weaker. The addition of such elements as chromium, silicon, and nickel produces steels with considerably higher strengths. Though frequently quite useful, these steels, however, are appreciably more expensive and often are not as easy to fabricate. A typical stress-strain diagram for a brittle steel is shown in Fig. 2. Unfortunately, low ductility or brittleness is a property usually associated with high strengths in steels Fig. 2 Typical stress-strain diagram for a brittle steel (although not entirely confined to high-strength steels). As it is desirable to have both high strength and ductility, the designer may have to decide between the two extremes or to compromise. A brittle steel may fail suddenly without warning when overstressed, and during erection could possibly fail due to the shock of erection procedures.

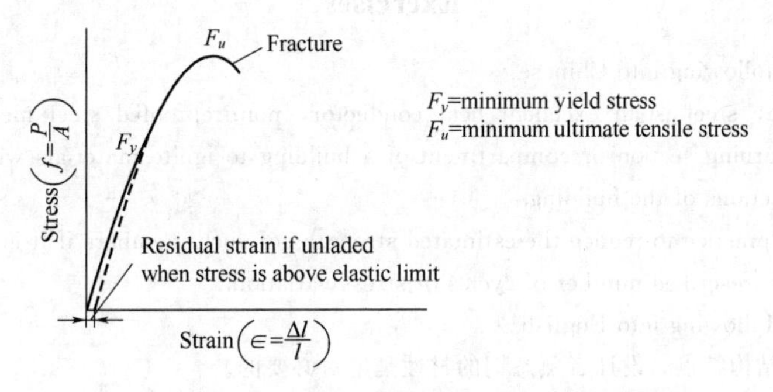

Fig. 2 Typical stress-strain diagram for a brittle steel

Brittle steels have a considerable range where stress is proportional to strain, but do not have clearly defined yield stresses. Yet, to apply many of the formulas given in structural steel design specifications, it is necessary to have definite yield stress values regardless of whether the steels are ductile or brittle.

If a mild steel member is strained beyond its elastic limit and then unloaded, it will not return to a condition of zero strain. As it is unloaded, its stress-strain diagram will follow a new path (shown by the dotted line in Fig. 2) parallel to the initial straight line. The result is a permanent or residual strain.

The yield stress for a brittle steel is usually defined as the stress at the point of unloading which corresponds to some arbitrarily defined residual strain (0.002 being the common value). In other words, we increase the strain by a designated amount and draw a line from that point, parallel to the straight-line portion of the stress-strain diagram, until the new line intersects the old. This intersection is the yield stress at that particular strain. If 0.002 is used, the intersection is usually referred to as the yield stress at 0.2 percent offset strain.

Words and Phrases

1	chromium	*n.*	铬
2	foundry	*n.*	铸造，铸件，铸造厂
3	incombustible	*a.*	防火的，不燃的
4	manganese	*n.*	锰
5	path	*n.*	路径，曲线，小径
6	phosphorus	*n.*	磷
7	residual	*a.*	剩余的，残留的
8	rolled	*a.*	辗压的，轧制的，辊轧的
9	scrap	*n.*	碎片，边角料
10	stiffen	*v.*	加强，加劲
11	stocky	*a.*	短粗的，结实的
12	uniformity	*n.*	均匀（性，度），一致，无变化
13	vitreous	*a.*	玻璃的，玻璃状（质）的
14	withstand	*v.*	抵抗，承受

Exercises

1. Translate the following into Chinese

(1) Furthermore, steel is an excellent heat conductor, nonfireproofed steel members may transmit enough heat from a burning section or compartment of a building to ignite materials with which they are in contact in adjoining sections of the building.

(2) The present practice to reduce the estimated strengths of such members if it is anticipated that they will have more than a prescribed number of cycles of stress variation.

2. Translate the following into English

(1) 为了了解钢结构特点，设计者熟悉钢的材性是绝对必要的。

(2) 不具备此性质的材料通常硬而脆，若承受突然荷载，很可能破坏。

Unit 13 Earthquake Prediction and Effect of Earthquake on Structures

In engineering design, it is important to estimate the likelihood of occurrence of a major earthquake within the useful lifetime of the structure (say, 100 years). It is not, of course, possible to predict exactly where and when a major event will occur. Thus, the term, "prediction", as used here only refers to the probability that a major earthquake can occur in the next, say, 100 years. This is the type of information needed in design. The likelihood of earthquake occurrence is often described in terms of a return period. A probability of 0. 01 for an earthquake intensity of 0. 3g at a site, for example, translates into average return period of 1/0. 01=100 years. Thus, if a building is designed for 0. 3g base acceleration, it is likely to experience this intensity once in the next 100 years.

Earthquake prediction is based on the past occurence of earthquake in a given region. Even though they occur randomly in time, earthquake of similar sizes (eg, approximately the same magnitudes) are expected to follow a trend in time and happen at relatively the same tine intervals. A simplified explanation for this can be found in the nature of tectonic earthquakes. As mentioned earlier, an earthquake occurs when stored energy releases. This happens when the resistance capacity of the fault zone is exceeded by the accumulated energy.

After release of energy during an earthquake, it will take nearly the same time for the energy to build up again and cause another earthquake. Unfortunately, reliable earthquake records in most areas are only available (at most) for the past few hundred years. Thus, it becomes difficult to predict the occurrence of the next major earthquake for a given area with a relatively high confidence level. Nevertheless, to determine the probable earthquake acceleration at a given site, the statistic of previous earthquakes of all magnitudes can be used along with proper modeling of the geometry of an existing fault and its location with respect to the site. This information is still valuable in design because an earthquake resistant structure is designed to withstand the maximum probable ground acceleration level that may occur during the lifetime of the structure.

Methods to determine the probability that certain ground acceleration levels will be experienced in a given site are referred to as "seismic risk" or "seismic hazard" analyses. These methods often utilize an earthquake prediction model along with an appropriate geometrical treatment of the fault area and potentials for rupture initiation on the fault. Furthermore, the uncertainty associated with the magnitudes of further earthquakes, depth of the ruptured area, and the location of point of the rupture initiation on the fault are considered in the analyses. These uncertainties result in a probability level that is a measure of the risk inherent in the design. Several seismic risk (or seismic hazard) analyses methods are currently available.

The difference between these methods is in how the geometry of the fault and rupture initiations are modeled and in the statistical model used to predict the future earthquakes. Seismic risk models often generate information in the form of curves. (shown in Fig. 1). These curves represent different ground acceleration levels versus their respective probabilities of occurrence or return periods for a given earthquake zone. Each curve belongs to a different site locate near the earthquake zone. By selecting a specific intensity level, the corresponding probability (risk) can be found from these curves and then used in design. Using curves such as those in Fig. 1, design maps can be developed for an earthquake zone. Such maps are often referred to as isoseismal or zoning maps and depict acceleration levels predicted for different locations in an earthquake zone. Seismic risk models are especially useful to develop isoseismal maps in areas where the earthquake data are only partially available.

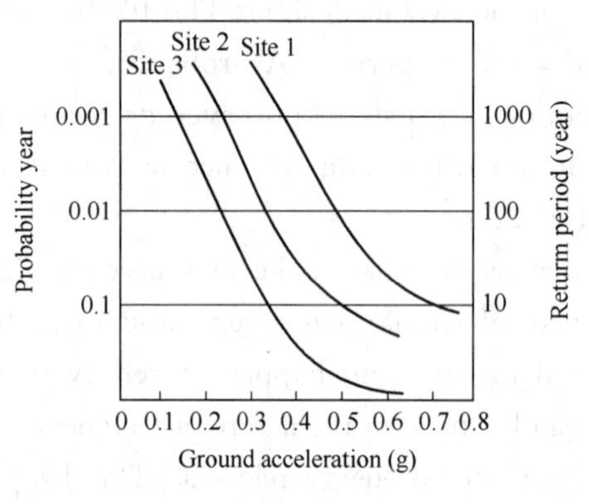

Fig. 1

Earthquake damage to a structure depends mainly on the response of the structure to the dynamic forces arising from the ground shading. The response of the structure depends on many factors including the ground motion peak value as well as its frequency and duration. Also the natural frequency of the building, soil condition at the site of the building, construction practice, and overall design of the building influence the building's response to the earthquake loads. Natural frequency of the building, in turn, depends on its geometry and the types of construction material used. Because of the variety of factors involved in the response of a building to the earthquake loads, it is difficult to generalize the type of damage that may occur to the building. Rather, the potential earthquake damage has to be investigated on a case-by-case basis. In this unit, a very broad discussion on the earthquake damage to the three classes of structure is presented. Specific information on the damage to each class should, however, be obtained from reports of post-earthquake damage evaluations and publication by organizations such as the Earthquake Engineering Research Institute (EERI). The three classes of structures considered in the following discussion are, namely, buildings, lifelines (telephone lines, gas mains, water mains to fire hydrants, etc), and inner-building utility piping systems.

Buildings

Damage to buildings varies to a great extent with the building's ability to dissipate energy and dampen the vibration during an earthquake. As described earlier, the natural frequency of a building and the frequency and duration of the ground shading also play important roles in the damage. The building's geometry and structural material influence its ductility, damping ability, and natural frequency of vibration. Thus, depending on the type of building, the damage can range from only a few cracks to major cracking and/or total collapse of the structure. Buildings with high natural frequency of vibration (i. e., short periods of vibration) exhibit much stiffer behavior during an earthquake. This is especially true for low-rise buildings (such as residential buildings, shopping centers, etc.) which often posses a natural period of vibration of 0. 006 to 0. 25 sec. The stiffer behavior may cause the building to suffer more damage during an earthquake. In such a case, however, if the ductility (i. e., the ability of the structure to deform as the load applies) is increased, the building becomes less stiff and thus less vulnerable to potential earthquake damage. When the natural period of the building is short, the vibration tends to quickly reach the stationary stage. As a result, the damage to the building will primarily depend on the duration of the ground shaking. With a longer duration, the building will be more likely to suffer damage. This is usually the case. (However, in the Mexico City earthquake of 1985, the more rigid buildings sustained less damage). Buildings with higher ductility exhibit longer periods of vibration. As a result, they are less likely to suffer damage. This especially true if the natural period of vibration of the building is much longer than the period of the ground vibration. In terms of individual structural members, a member with more ductility deforms more and thus delays reaching the critical failure stage. This is the type of behavior in metal structures and, to certain extent, in timber structures. However, concrete and masonry structures are less ductile and need to be designed with an appropriate amount of reinforcement and special reinforcement detailing at the beam-column connections so that they can maintain the needed ductility in an earthquake load environment.

Lifelines

Damage to lifelines occurs in the form of dislocation and/or rupture in water mains, gas mains, and sewer lines, disruption of telephone lines, severe cracking in roadways, and collapse of bridges. A dramatic example of this occurred in San Francisco's Marina District during the Loma Prieta earthquake in 1989. Damage to water mains may cause flooding; and gas main breaks can cause fire: The 1923 earthquake in Tokyo resulted in destructive fires over 40 percent of the city. Underground pipeline systems have been extensively damaged in severe earthquakes such as Long Beach in 1973, Alaska in 1964, and San Fernando in 1971. Following the May 16, 1968 Todachi-oki earthquake magnitude, 7. 9 on Richter scale in Japan, 947 cases of leaks were discovered in the water distribution system of the city Hachinoche. The damage to lifelines is especially critical because essential services (which are vitally needed in an emergency situation) are drastically affected.

During an earthquake, a lifeline system may be damaged (a) due to severe system vibration developed as a result of the ground shaking, and (b) due to the fault-rupture running through the lifeline. The latter happens if the earthquake is of the type with a shallow focus so that the ruptured area can hit the ground surface. The vibration induced in pipes can cause either the pipe to break or can cause joints to become loose and thus develop leaks (gas or water).

Inner-Building Piping System

Damage to interior piping is especially critical to gas pipes where gas leaks may develop. The post earthquake evaluation of the October 1, 1987 Whittier Narrows earthquake revealed some 1400 cases of gas leaks. However, only a very small portion of these leaks ended up as sizeable fires. Similar types of damage were also observed in the 1971 San Fernando earthquake.

Words and Phrases

1	damper	*n.*	阻尼（减速，减振）器
2	damp down		减轻，缓冲
3	dissipative	*a.*	消（分，扩）散的，散逸的
4	response	*n.*	应（回）答，反（响）应（曲线），特性曲线
5	vibrate	*vt*，*vi.*	振荡（动），摇摆
6	decay	*vt*，*vi.*，*n.*	腐朽（化），衰变（退，减）
7	time-varying	*a.*	随时间变化的
8	motion	*n.*	运（移）动，运行（转）（机械）
9	rebound	*vi.*	弹回，回跳
10	rebind	*vt.*	重捆，重新装（包扎）
11	gust	*n.*	（一）阵（骤）风（雨）
12	continuum	*n.*	连续（体）
13	truck	*n.*，*vt.*	（装上）卡车，货车
14	liberalize	*vt.*，*vi.*	放宽（限制）范围，使自由化
15	liberal	*a.*	开明的，（思想）自由开朗的
16	sewer	*n.*	下水道
17	quadrilateral	*a.*，*n.*	四边（角）形，四方面的
18	hyperbolic	*a.*	双曲线的，夸大的
19	hyperbola	*n.*	双曲线
20	parabolic	*a.*，*n.*	抛物面（线，体）的
21	variation	*n.*	（可）变量（数，项）
22	time history		随时间的变化，时间关系曲线图

Exercises

1. Translate the following into Chinese

(1) Even though they occur randomly in time，earthquake of similar sizes are expected to follow a trend in time and happen at relatively the same time intervals.

(2) Because of the variety of factors involved in the response of building to the earthquake loads，it is difficult to generalize the type of damage that may occur to the building.

2. Translate the following into English

(1) 地震预测以该地区过去发生的地震为基础。

(2) 地震对结构的破坏主要取决于结构对地面震动产生的力的反应。

Unit 14 Computer-Aided Drafting and Design

CADD is an acronym for computer-aided drafting and design. Initially, the letters CAD referred to computer-aided (or assisted) drafting, but now they can mean computer-aided drafting, computer-aided design, or both. ❶ CADD (with two Ds) better describes current technology since many popular software packages include both two-dimensional drafting and three-dimensional design functions. In addition, many CADD systems provide analysis capabilities, schedule production, and reporting. All are part of the design process.

CADD has dramatically changed the way that designers develop and record their ideas. Even though the principles of graphic communication have not changed, the method of creating, manipulating and recording design concepts has changed with computers. ❷ The techniques of drafting, organization of views, projection, representation of elements in a design, dimensioning, etc. are the same. Yet, drawing boards, triangles, scales, and other traditional drafting equipment are no longer required to communicate a design idea. A designer/drafter using a computer system and the appropriate software can:

- Plan a part, structure, or whatever product is needed.
- Modify the design without having to redraw the entire plan.
- Call up symbols or base drawings from computer storage.
- Automatically duplicate forms and shapes commonly used.
- Produce schedules or analyses.
- Produce hardcopies of complete drawing or drawing elements in a matter of minutes.

Just a few short years ago CADD programs were crude and clumsy, not user friendly, and difficult to use. Most of the early packages required a mainframe computer or powerful minicomputer to handle the calculations. ❸ Even then, they were extremely slow. The cost of such a system was so high that only large companies or school systems could afford to purchase one.

Today, this has all changed. Some very large and complex applications still require mainframe, super-minicomputers, or minicomputers. Plus, these systems have become easier to use, faster, and even more powerful. However, microcomputer-based CADD (micro-CADD) systems have steadily improved. Now, they meet the needs of most users. These newer systems are packed with features. Their cost is only a small fraction of a mainframe system, making them affordable for almost every classroom and small business. The trend is toward smaller computers running specialized software programs.

Popular micro CADD systems offer high performance at an affordable price. Increased performance improves productivity and helps to produce more accurate drawings. All types of architectural, engineering, and construction (AEC) drawings may be produced with a CADD system.

Once a design has been completed and stored in the computer, it can be called up whenever needed for copies or revisions. ❹ Revising CADD drawings is where the true time savings is realized. Frequently, a revision that requires several hours to complete using traditional drafting methods can be done in a few minutes on a CADD system. In addition, some CADD packages automatically produce updated schedules after you revise the original plan.

Another benefit of CADD is symbol libraries. Inserting standard symbols and shapes is quick, easy, and accurate. Once a standard symbol has been drawn and stored in the library, it can be called up and placed as many times, in as many drawings, as required. For example, symbols for trees, furniture, doors and windows, and common appliances are usually included in an architectural symbol library. During insertion, the symbol may be scaled, rotated, or mirrored to meet a specific need. In addition, unique symbols may be designed and added to the library for local applications.

A CADD system generally consists of the following hardware: computer or processor, monitor, graphics adapter, input and pointing device, and hardcopy device. Each of these components serves a unique purpose in the system.

The heart of a CADD system is the computer. It serves as the center of activity since all information flows through it. Peripheral devices (plotters, digitizing tablets, monitors, etc.) are connected to the computer. Most CADD systems require computers that can process a large amount of information very rapidly. Processing speed is an important consideration when choosing a CADD system computer. The same computer may be used for non-CADD applications, such as word processing and accounting, or the computer may be specially "dedicated" to drafting and design functions.

The engine or "brain" of the computer system is the CPU (central processing unit). The CPU, also known as a microprocessor in microcomputers, is a computer chip that executes the software program and keeps the various system components synchronized and operating in harmony. Three basic computer CPUs are found in microcomputers today. They are rated according to the size of instruction they can process at one time—8, 16, and 32 bits. In general, the more powerful the CPU, the faster the computer.

The CPU stores data and programs in another hardware component, called Random Access Memory (RAM). Memory is for short-term storage and should not be confused with disk storage. The amount of memory is important when choosing CADD programs. Some require larger amounts of storage than your computer can handle. Microcomputers should have a minimum of 640k RAM for most CADD software. Once a computer is turned off, all the information contained in RAM is lost. This is why you store data on permanent storage devices such as floppy disks and hard disks.

Math coprocessors are "number crunches" that take some of the workload off the CPU. With CADD, the coprocessor speeds up the generation of on-screen graphics and is required to operate many CADD software programs. The execution of display commands such as Redraw, Pan, and Zoom are very slow if the computer does not have a math coprocessor. The advent of the math coprocessor is one main factor that led to the sudden popularity of CADD on microcomputers.

The math coprocessor speeds up processing because graphics information is stored as coordinate data. These data (numbers) require rapid calculation to be useful. The speed using a math coprocessor is about five times that of the same machine without this chip.

Connected to the computer are input and output devices such as a plotter, digitizing tablet, and monitor. These connect to ports on the back of the computer.

The monitor, or display, allows the computer to communicate with the operator. Monitors are available in a broad spectrum of sizes, prices, quality, and applications. Important considerations include: compatibility with the computer you are using, color or monochrome, physical size of the monitor, and character resolution.

The graphics monitor requires a special circuit board in the computer called a graphics adapter or controller. The adapter should be compatible with the resolution and color characteristics required by the monitor. This lets the monitor display its best color and resolution capabilities. Typical graphics adapters include: EGA (enhanced graphics adapter), CGA (color graphics adapter), MDA (monochrome display adapter) and VGA (video graphics array).

All CADD systems require one or more input devices. Typical input devices include: keyboard, mouse, digitizing tablet and puck or stylus, light pen, joystick, thumb wheels, and track ball.

The information created on a CADD system may be classified as softcopy or hardcopy. Softcopy includes user prompts, instructions, and a visual record of the operations as they are performed. This information is output to the monitor. Hardcopy output includes drawings, parts lists, bills of materials, and specifications. These are produced on paper, vellum, or film using a plotter or printer.

The hardcopy device you have depends on the intended use of the output. The most common devices used to produce hardcopy output include: pen plotters, laser printers, thermal plotters, electrostatic plotters, color impact printers, daisy wheel printers, and film recorders. When evaluating these devices, consider their accuracy, quality of lines and text, speed, plot size, and color possibilities.

Words and Phrases

1	acronym	*n.*	简称，只取首字母的缩写词
2	manipulate	*v.*	操纵，操作，生成
3	redraw	*v.*	重画，刷新屏幕
4	duplicate	*v.*	复写，复制，转录，重复
5	hardcopy	*v.*, *n.*	硬拷贝
6	insertion	*n.*	插入，嵌入，插页
7	hardware	*n.*	硬件，部件
8	processor	*n.*	处理机，处理器，处理程序
9	monitor	*n.*	监视器，监督程序
10	plotter	*n.*	绘图机，绘图仪，图形显示器
11	port	*n.*	断口，通讯口，进出口
12	monochrome	*n.*	单色，单色图像
13	compatibility	*n.*	兼容性，相容性，适用性，互换性

14	video	*n*.	试品，影响，电视（图像）
15	stylus	*n*.	唱针，铁壁（汉字输入用），指示笔
16	keyboard	*n*.	键盘，用键盘写入
17	joystick	*n*.	控制杆，操纵杆，游戏杆
18	prompt	*n*.，*v*.	提示符，提示
19	vellum	*n*.	精制犊皮纸，牛皮纸
20	mainframe	*n*.	主机，大型机
21	adapter	*n*.	适配器，转换接头，附件
22	digitize	*v*.	数字化，将资料数字化
23	computer-aided		计算机辅助
24	Random Access Memory		随机存取存储器

Notes

❶ Initially，the letters CAD referred to computer-aided（or assisted）drafting，but now they can mean computer-aided drafting，computer-aided design，or both.

起初，CAD三个字母是指计算机辅助制图，但现在它们可以指计算机辅助制图，或者计算机辅助设计，或者计算机辅助制图与设计。

❷ Even though the principles of graphic communication have not changed，the method of creating，manipulating，and recording design concepts has changed with computers.

即使图形表达的原则没有变化，创建、处理和记录设计概念的方法却因为计算机而改变了。

❸ Most of the early packages required a mainframe computer or powerful minicomputer to handle the calculations.

大多数早期的软件包都需要大型计算机或大功率的小型计算机来处理计算。

❹ Once a design has been completed and stored in the computer，it can be called up whenever needed for copies or revisions.

设计图一旦完成并存储在计算机里，则无论何时需要拷贝或修改都能够调用。

Exercises

1. Translate the following into Chinese

（1）Revising CADD drawings is where the true time savings is realized. Frequently，a revision that requires several hours to complete using traditional drafting methods can be done in a few minutes on a CADD system. In addition，some CADD packages automatically produce updated schedules after you revise the original plan.

（2）The math coprocessor speeds up processing because graphics information is stored as coordinate data，these data（numbers）require rapid calculation to be useful，the speed using a math coprocessor is about five times that of the same machine without this chip.

2. Translate the following into English

（1）尽管设计者沟通的原则没有改变，但生成、操作和记录设计概念的方法已经通过计算机发生变化。

（2）一旦标准样本被绘制并储存在计算机图库中，它可以在许多时候被调出并放在许多图纸需要的地方。

Unit 15 Construction Engineering

Construction engineering is a specialized branch of civil engineering concerned with the planning, execution, and control of construction operations for such projects as highways, buildings, dams, airports, and utility lines.

Planning consists of scheduling the work to be done and selecting the most suitable construction methods and equipment for the project. Execution requires the timely mobilization of all drawings, layouts, and materials on the job to prevent delays to the work. Control consists of analyzing progress and cost to ensure that the project will be done on schedule and with the estimated cost.

Planning. The planning phase starts with a detailed study of construction plans and specifications. From this study a list of all items of work is prepared, and related items are then grouped together for listing on a master schedule. A sequence of construction and the time to be allotted for each item is then indicated. The method of operation and the equipment to be used for the individual work items are selected to satisfy the schedule and the character of the project at the lowest possible cost.

The amount of time allotted for a certain operation and the selection of methods of operation and equipment that is readily available to the contractor. After the master or general construction schedule has been drawn up, subsidiary detailed schedules or forecasts are prepared from the master schedule. These include individual schedules for procurement of material, equipment, and labor, as well as forecasts of cost and income.

Execution. The speedy execution of the project requires the ready supply of all materials, equipment, and labor when needed. The construction engineer is generally responsible for initiating the purchase of most construction materials and expediting their delivery to the project. Some materials, such as structural steel and mechanical equipment, require partial or complete fabrication by a supplier. For this fabricated materials the engineer must prepare or check all fabrication drawings for accuracy and case of assembly and often inspect the supplier's fabrication.

Other construction engineering duties are the layout of the work by surveying methods, the preparation of detail drawings to clarify the design engineer's drawings for the construction crews, and the inspection of the work to ensure that it complies with plans and specifications.

On most large projects it is necessary to design and prepare construction drawings for temporary construction facilities, such as drainage structures, access roads, office and storage buildings, formwork, and cofferdams. Other problems are the selection of electrical and mechanical equipment and the design of structural features for concrete material processing and mixing plants and for compressed air, water, and electrical distribution systems. ❶

Control. Progress control is obtained by comparing actual performance on the work against the desired performance set up on the master or detailed schedules. ❷ Since delay on one feature of the project could easily affect the entire job, it is often necessary to add equipment or crews to speed up the work.

Cost control is obtained by comparing actual unit costs for individual work items against estimated or budgeted unit costs, which are set up at the beginning of the work. A unit cost is obtained by dividing the total cost of an operation by the number of units in that operation.

Typical units are cubic yards for excavation or concrete work and tons for structural steel. The actual unit cost for any item at any time is obtained by dividing the accumulated costs charged to that item by the accumulated units of work performed.

Individual work items costs are obtained by periodically distributing job costs, such as payroll and invoices to the various work item accounts. Payroll and equipment rental charges are distributed with the aid of time cards prepared by crew foreman. The cards indicate the time spent by the job crews and equipment on the different elements of the work. The allocation of material costs is based on the quantity of each type of material used for each specific item.

When the comparison of actual and estimated unit costs indicates an overrun; an analysis is made to pinpoint the cause. If the overrun is in equipment costs, it may be that the equipment has insufficient capacity or that it is not working properly. If the overrun is in labor costs, it may be that the crews have too many men, lack of proper supervision, or are being delayed for lack of materials or layout. In such cases time studies are invaluable in analyzing productivity.

Construction operations are generally classified according to specialized fields. These include preparation of the project site, earthmoving, foundation treatment, steel erection, concrete placement, asphalt paving, and electrical and mechanical installations. Procedures for each of these fields are generally the same, even when applied to different projects, such as buildings, dams, or airports. However, the relative importance of each field is not the same in all cases. For a description of tunnel construction, which involves different procedures.

Preparation of site. This consists of the removal and clearing of all surface structures and growth from the site of the proposed structure. A bulldozer is used for small structures and trees. Larger structures must be dismantled.

Earthmoving. This includes excavation and the placement of earth fill. Excavation follows preparation of the site, and is performed when the existing grade must be brought down to a new elevation. Excavation generally starts with the separate stripping of the organic topsoil, which is later reused for landscaping around the new structure. This also prevents contamination of the nonorganic material which may be required for fill. Excavation may be done by any of several excavators, such as shovels, draglines, clamshells, cranes, and scrapers.

Efficient excavation on land requires a dry excavation area, because many soils are unstable when wet and cannot support excavating and hauling equipment. Dewatering becomes a major operation when the excavation lies below the natural water table and intercepts the groundwater flow. When this occurs, dewatering and stabilizing of the soil may be accomplished by wellpoints and electroosmosis.

Some materials, such as rock, cemented gravels, and hard clays, require blasting to loosen or fragment the material. Blast holes are drilled in the material; explosives are then placed in the blast holes and detonated. The quantity of explosives and the blast-hole spacing are dependent upon the type and structure of the rock and diameter and depth of the blast holes.

After placement of the earth fill, it is almost always compacted to prevent subsequent settlement. Compaction is generally done with sheep's-foot, grid, pneumatic-tired, and vibratory-type rollers, which are towed by tractors over the fill as it is being placed. Hand-held, gasoline-driven rammers are used for compaction close to structures where there is no room for rollers to operate.

Foundation treatment. Where subsurface investigation reveals structural defects in the foundation area to be used for a structure, the foundation must be strengthened. Water passages, cavities, fissures, faults, and other defects are filled and strengthened by grouting. Grouting consists of injection of fluid mixtures under pressure. The fluids subsequently solidify in the voids of the strata. Most grouting is done with cement and water mixtures but other mixture ingredients are asphalt, cement and clay, and precipitating chemicals.

Steel erection. The construction of a steel structure consists of the assembly at the site of mill-rolled or shop-fabricated steel sections. The steel sections may consist of beams, columns, or small trusses which are joined together by riveting, bolting, or welding. It is more economical to assemble sections of the structure at a fabrication shop rather than in the field, but the size of preassembled units is limited by the capacity of transportation and erection equipment. The crane is the most common type of erection equipment, but when a structure is too high or extensive in area to be erected by a crane, it is necessary to place one or more derricks on the structure to handle the steel. In high structures the derrick must be constantly dismantled and reerected to successively higher levels to raise the structure. For river bridges, the steel may be handled by cranes on barges, or, if the bridges is too high, by traveling derricks which ride on the bridge being erected. Cables for long suspension bridges are assembled in place by special equipment that pulls the wire from a reel, set up at one anchorage, across to the opposite anchorage, repeating the operation until the bundle of wires is of the required size.

Concrete construction. Concrete construction consists of several operations: forming, concrete production, placement, and curing. Forming is required to contain and support the fluid concrete within its desired final outline until it solidifies and can support itself. The form is made of timber or steel sections or a combination of both and is held together during the concrete placing by external bracing or internal ties. The forms and ties are designed to withstand the temporary fluid pressure of the concrete.

The usual practice for vertical walls is to leave the forms in position for at least a day after the concrete is placed. They are removed when the concrete has solidified or set. Slip-forming is a method where the form is constantly in motion, just ahead of the level of fresh concrete. The form is lifted upward by means of jacks which are mounted on vertical rods embedded in the concrete and are spaced along the perimeter of the structure. Slip forms are used for high structures such as silos, tank, or chimneys.

Concrete may be obtained from commercial batch plants which deliver it in mix trucks if the job is close to such a plant, or it may be produced at the job site. [3] Concrete production at the job site requires the erection of a mixing plant, and of cement and aggregate receiving and handling plants. Aggregates are sometimes produced at or near the job site. This requires opening a quarry and erecting processing equipment such as crushers and screens.

Concrete is placed by chuting directly from the mix truck, where possible, or from buckets handled by means of cranes or cableways, or it can be pumped into place by special concrete pumps.

Curing of expose surfaces is required to prevent evaporation of mix water or to replace moisture that does evaporate. The proper balance of water and cement is required to develop full design strength.

Concrete paving for airports and highways is a fully mechanized operation. Batches of concrete are placed between the road forms from a mix truck or a movable paver, which is a combination mixer and placer. A series of specialized pieces of equipment, which ride on the forms, follow to spread and vibrate the concrete, smooth its surface, cut contraction joints, and apply a curing compound.

Asphalt paving. This is an amalgam of crushed aggregate and a bituminous binder. It may be placed on the roadbed in separate operations or mixed in a mix plant and spread at one time on the roadbed. Then the pavement is compacted by rollers.

Words and Phrases

1	execution	*n.*	执行，完成，实施，施工
2	master	*a.*	主要的，总的，熟练的
3	clarify	*v.*	澄清，阐明，净化，解释
4	budget	*n.*，*v.*	预算，作预算，编入预算
5	payroll	*n.*	薪水册，发放工资额，工资单
6	invoice	*n.*	发票，发货单，开发票，记清单
7	overrun	*v.*	超过，超出，超限
8	productivity	*n.*	生产力，劳动生产率
9	bulldozer	*n.*	推土机，开土机，压路机
10	dismantle	*v.*	拆除，拆卸，粉碎
11	shovel	*n.*	铲，挖掘机，单斗挖土机
12	dragline	*n.*	拉索，拉铲挖土机
13	clamshell	*n.*	抓斗，蛤壳式挖泥机
14	scraper	*n.*	铲运机，刮土机，平土机
15	haul	*v.*	拖，拉，用力拖拉，拖运
16	subsidiary	*a.*	辅助的，补充的，次要的
17	contamination	*n.*	污染，污染物
18	wellpoint	*n.*	降低地下水位的井点，深坑点
19	fragment	*n.*	断片，碎块，凝固
20	solidify	*v.*	固化，固结，凝固

21	ingredient	*n.*	组分，成分，配料
22	silo	*n.*	筒仓，竖井，(导弹)发射井
23	amalgam	*n.*	混合物，软的混合物
24	slip form	*n.*	滑动模板，滑模（施工法）

Notes

❶ Other problems are the selection of electrical and mechanical equipment and the design of structural features for concrete material processing and mixing plants and for compressed air, water, and electrical distribution systems.

其他的问题是选择电气与机械设备和具体设计混凝土原料加工与搅拌厂以及压缩空气，配水，配电系统。

❷ Progress control is obtained by comparing actual performance on the work against the desired performance set up on the master or detailed schedules.

进度控制是通过比较工程实际进度与主要（或详细）进度表中确定的预期进度来进行的。

❸ Concrete may be obtained from commercial batch plants which deliver it in mix trucks if the job is close to such a plant, or it may be produced at the job site.

如果工地离商品混凝土搅拌厂很近，混凝土就可以从用搅拌车运送混凝土的商品混凝土搅拌厂获得，否则可以在工地制作。

Exercises

1. Translate the following into Chinese

(1) The construction engineer is generally responsible for initiating the purchase of most construction materials and expediting their delivery to the project. Some materials, such as structural steel and mechanical equipment, require partial or complete fabrication by a supplier.

(2) When the comparison of actual and estimated unit costs indicates an overrun; an analysis is made to pinpoint the cause. If the overrun is in equipment costs, it may be that the equipment has insufficient capacity or that it is not working properly. If the overrun is in labor costs, it may be that the crews have too many men, lack of proper supervision, or are being delayed for lack of materials or layout.

(3) The steel sections may consist of beams, columns, or small trusses which are joined together by riveting, bolting, or welding. It is more economical to assemble sections of the structure at a fabrication shop rather than in the field, but the size of preassembled units is limited by the capacity of transportation and erection equipment.

2. Translate the following into English

(1) 成本控制通过比较单项工程实际单位成本和工程开始时制定的单位成本的估算或预算来实现。

(2) 当挖土位于自然水位以下并截断地面水流时，降水成为一个主要的作业。

(3) 在地质勘察显示供结构使用的区域地基有缺陷的地方，基础必须加强。

Unit 16 Civil Engineering Contracts

A simple contract consists of an agreement entered into by two or more parties, whereby[1] one of the parties undertakes to do something in return for something to be undertaken by the other. A contract has defined as an agreement which directly creates and contemplates an obligation. The word is derived from the Latin *contractum*, meaning drawn together.

We all enter into contracts almost every day for the supply of goods, transportation and similar services, and in all these instances we are quite willing to pay for the services we receive. Our needs in these cases are comparatively simple and we do not need to enter into lengthy or complicated negotiations and no written contract is normally executed. Nevertheless, each party to the contract has agreed to do something, and is liable for breach of contract if he fails to perform his part of the agreement.

In general, English law requires no special formalities in making contract but, for various reasons, some contracts must be made in a particular form to be enforceable and, if they are not made in that special way, then they will be ineffective. Notable among these contracts are contracts for the sale and disposal of land,[2] and "land", for this purpose, includes anything built on the land, as, for example, roads, bridges and other structures.

It is sufficient in order to create a legally binding contract, if the parties express their agreement and intention to enter into such a contract.[3] If, however, there is no written agreement and a dispute arises in respect of the contract, then the Court that decides the dispute will need to ascertain the terms of the contract from the evidence given by the parties, before it can make a decision on the matters in dispute.

On the other hand if the contract terms are set out in writing in document, which the parties subsequently sign, then both parties are bound by these terms even if they do not read them. Once a person had signed a document he is assumed to have read and approved its contents, and will not be able to argue that the document fails to set out correctly the obligations which he actually agreed to perform. Thus by setting down the terms of a contract in writing one secures the double advantage of affording evidence and avoiding disputes.

The law relating to contracts imposes on each party to a contract a legal obligation to perform or observe the terms of the contract, and gives to the other party the right to enforce the fulfillment of these terms or to claim "damages" in respect of the loss sustained in consequence of the breach of contract.

Most contracts entered into between civil engineering contractors and their employers are of the type known as "entire" contracts. These are contracts in which the agreement is for special works to be undertaken by the contractor and no payment is due until the work is complete.

In an entire contract, where the employer agrees to pay a certain sum in return for civil engineering work, which is to be executed by the contractor, the contractor is not entitled to any payment if he abandons the work prior to completion, and will be liable in damages for breach of contract. Where the work is abandoned at the request of employer, or results from circumstances that were clearly foreseen when the contract was entered into and provided for in its terms, then the contractor will be paid as much as he has earned.

It is, accordingly, in the employer's interest that all contracts for civil engineering work should be entire contracts to avoid the possibility of work being abandoned prior to completion. However, contractors are usually unwilling to enter into any contracts, other than the very smallest, unless provision is made for interim payments to them as the work proceeds. For this reason the standard form of civil engineering contract provides for the issue of interim certificates at various stages of the works.

It is customary for the contract further to provide that a prescribed proportion of the sum due to the contractor on the issue of a certificate shall be withheld. This sum is known as "retention money" and serves to insure the employer against any defects that may arise in the work. The contract does, however, remain an entire contract, and the contractor is not entitled to receive payment in full until the work is satisfactorily completed, the maintenance period expired and the maintenance certificate issued.

That works must be completed to the satisfaction of the employer, or his representative, does not give the employer the right to demand an unusually high standard of quality throughout the works, in the absence of a prior express agreement. Otherwise the employer might be able to postpone indefinitely his liability to pay for the works. The employer is normally only entitled to expect a standard of work that would be regarded as reasonable by competent persons with considerable experience in the class of work covered by the particular contract. The detailed requirements of the specification will have a considerable bearing on these matters.

The employer or promoter of civil engineering works normally determines the conditions of contract, which define the obligation and performances to which the contractor will be subject. He often selects the contractor for the project by some form of competitive tendering and, any contractor who submits a successful tender and subsequently enters into a contract is deemed in law to have voluntarily accepted the conditions of contract adopted by the promoter.

The obligations that a contractor accepts when he submits a tender are determined by the form of invitation to tender. In most cases the tender may be withdrawn at anytime until it has been accepted and may, even then, be withdrawn if the acceptance is stated by the promoter to be subject to formal contract as is often the case.

The employer does not usually bind himself to accept the lowest or indeed any tender and this is often stated in the advertisement. A tender is, however, normally required to be a definite offer and acceptance of it gives rise legally to a binding contract.

A variety of contractual arrangements are available and engineer will often need to carefully select the form of contract which is best suited for the particular project. The employer is entitled to know the reasoning underlying the engineer's choice of contract.

Types of contract are virtually classified by their payment system: (1) price-based: lump sum and admeasurements (prices or rates are submitted by the contractor in his tender); and (2) cost-based: cost-reimbursable and target cost (the actual costs incurred by the contractor are reimbursed, together with a fee for overheads and profit).

Words and Phrases

1	contract	*n.*	合同，契约
2	obligation	*n.*	义务，责任
3	(be) liable for		对……负责的
4	breach	*n.*, *v.*	破坏，违反
5	withhold	*v.*	扣留
6	promoter	*n.*	发包者
7	tender	*n.*, *v.*	标书；招投标
8	overhead	*n.*	管理费
9	profit	*n.*	利润
10	breach of contract		违约
11	binding contract		有（法律）约束力的合同
12	contract terms		合同条款
13	claim "damages"		索赔
14	interim payment (certificates)		中间付款（验收，或证书）
15	retention money		保留（滞留金）
16	lump sum contract		总价合同
17	admeasurement contract		计价合同
18	cost-reimbursable contract		成本补偿合同
19	target cost contract		目标成本合同

Notes

❶ whereby = by which，意为"借此"，引出一个非限定性定语从句。

❷ 为强调表语，采用倒装语句，把表语 Notable 挪到句首，意为"这些合同中，值得注意的是……"。

❸ It is sufficient in order to create... 句中，It 为形式主语，不定式 to create a legally binding contract 为真实主语，介词短语 in order 的含义是"符合要求，妥当"，充当状语，修饰 sufficient。

Exercises

1. Translate the following into Chinese

(1) The contractor is not entitled to receive payment in full until the work is satisfactorily completed, the maintenance period expired and the maintenance certificate issued.

(2) The employer does not usually bind himself to accept the lowest or indeed any tender and this id often stated in the advertisement.

2. Translate the following into English

(1) 最终，由法院根据合同条款和各方证词来裁决争议。

(2) 若承包商不履行维修期的合同条款，业主有权拒付保留金。

Unit 17 Composite Structures and Design Philosophy

Composite Beams and Slabs

The design of structures for buildings and bridges is mainly concerned with the provision and support of load-bearing horizontal surfaces. Except in long-span bridges, these floors or decks are usually made of reinforced concrete, for no other material has a better combination of low cost, high strength, and resistance to corrosion, abrasion, and fire.

The economical span for a reinforced concrete slab is little more than that at which its thickness becomes just sufficient to resist the point loads to which it may be subjected or, in buildings, to provide the sound insulation required. For spans of more than a few metres it is cheaper to support the slab on beams or walls than to thicken it. When the beams are also of concrete, the monolithic nature of the construction makes it possible for a substantial breadth of slab to act is the top flange of the beam that supports it.

At spans of more than about 10 m, and particularly where the susceptibility of steel to damage by fire is not a problem, as for example in bridges and multi-storey car parks, steel beams become cheaper than concrete beams. It used to be customary to design the steelwork to carry the whole weight of the concrete slab and its loading; but by about 1950 the development of shear connectors had made it practicable to connect the slab to the beam, and so to obtain the T-beam action that had long been used in concrete construction. The term "composite beam" as used in this book refers to this type of structure.

The same term is used for beams in which prestressed and in-situ concrete act together, and there are many other examples of composite action in structures, such as between brick walls and the beams supporting them, or between a steel-framed shed and its cladding; but these are outside the scope of this book.

No income is received from money invested in the construction of a multi-storey building such as a large office block until the building is occupied. For a construction time of two years, this loss of income from capital may be 15% of the total cost of the building; that is, it may be over half the total cost of the structure. The construction time is strongly influenced by the time taken to construct a typical floor of the building, and here structural steel has an advantage over in-situ concrete. Even more time can be saved if the floor slabs are cast on permanent formwork that also acts as bottom reinforcement. These composite slabs, using corrugated steel sheeting as the formwork, have long been used in tall buildings in North America. In Europe, precast concrete permanent formwork and full-thickness precast floor slabs have also been found to be economical, particularly in multi-storey car parks.

At a time of rapid inflation and high interest rates, no comparison of the relative costs of different types of structure is meaningful unless account is taken of construction times. An example of this is provided by a 29-storey office block that was completed in London in 1974. The structure consist of a slip-formed reinforced concrete core surrounded by a steel frame.

This was designed to act compositely with the floor slabs, for which precast concrete planks and in-situ topping was used. On paper, this structure was more expensive than a reinforced concrete frame, but its construction time of 27 months was 8 months less than that for the concrete alternative. The consequent reduction in the total of the scheme justified the adoption of the composite structure.

The degree of fire that must be provided is another factor that influences the choice between concrete, composite, and steel structures, and here concrete has an advantage. Little or no fire protection is required for multi-storey car parks, a moderate amount for office blocks, more for public buildings, and most of all for multi-storey warehouses. Many methods have been developed for providing steel work with fire protection, and information on them is readily available. Encasement in concrete is an economical method for steel columns, since the casing provides a substantial gain in strength.[1] Most of the methods used for composite beams rely on lightweight materials, but in a few jobs the steel beams have been encased in concrete before erection. The encasement, contributes little to the strength of the beam, but requires light-gauge reinforcement to control the width of cracks and to hold it in place during a fire.

The choice between steel, concrete and composite construction for a particular structure thus depends on many factors that are outside the scope of this book. But it is clear from recent practice that composite construction is particularly competitive in medium or long span structures where a concrete slab is needed for other reasons, where fire protection of steelwork is not required, and where there is a premium on rapid construction.

Composite Columns and Frames

When the stanchions in steel frames were first cased in concrete to protect them from fire, they were still designed for the applied load as if uncased. Then engineers realized that the encasement reduced the effective slenderness of a stanchion, and so increased its buckling load. Empirical methods for calculating the reduced slenderness are now given in design specifications for structural steelwork.

This approach, although simple, is not rational, for the concrete encasement also carries its share of the load. Methods are now available for designing cased stanchions as composite columns. These take account of the true behavior of these members, as found in tests to failure.

When fire protection for the steel is not required, a composite column can be constructed without the use of formwork by filling a steel tube with concrete. Research on filled tubes led to their use in 1966 in a four-level motorway interchange at Almondsbury, near Bristol.

In framed structures, there may be composite beams, composite columns, or both. In developing design methods that take account of the interaction between beams and columns, it is necessary to consider many types of beam-column joint, and also to reconcile the differences between the methods now in use for concrete frames and for steel frames.

Two buildings with rigid-jointed composite frames were built in Great Britain in the early 1960s, at Cambridge and London. In a sense, these were full-scale experiments, for university staff engaged in relevant research were involved in their design. Also, at Imperial College, London, composite columns were used in the lower half of an otherwise reinforced concrete-framed structure of about 15 storeys, in order to maintain a constant size of column over the whole height of the building; but no account of this design has been published.

Design Philosophy

An essential part of the design process is to take account of the random nature of loading, the variability of materials, and the defects that occur in construction, in such a way that the probability of unserviceability or failure of the structure during its design life is reduced to an acceptably low level. Extensive study of this subject since about 1950 has led to the incorporation of the older "safety factor" and "load factor" design methods into a more comprehensive "limit state" design philosophy. Its first important application in Great Britain was in Code of Practice 110, *The Structural use of Concrete*, published in 1972.

It is the policy of the British Standards Institution that future codes of practice and design specifications for structures shall be in limit-state form. Work began in 1968 on a "Unified Bridge Code", to include steel, concrete, and composite superstructures and columns for bridges of all spans. The revision of BS 449 for steel structures in building began in 1969, and in 1970 it was decided to include composite steel-concrete structures within the new Standard, and to discontinue work on CP 117. The completion of the Bridge Code has been delayed by the need to take account in the "steel" section of the lessons learned from several failures of box girders during the period 1968 to 1972, and from the research that followed. This subject is discussed further in Volume 2. It is hoped that the section of the new BS 449 on composite structures will be issued for public comment during 1976.

Limit-state design philosophy is used throughout this book, as this enables the principles and assumptions that underlie design methods to be set out more clearly than is possible in terms of the older philosophies. Many accounts of limit-state philosophy are available. The following brief summary of it is intended only to relate it to the older methods and to assist the reader to follow the design examples.

In traditional "elastic" design, "working" or "permissible" stresses are obtained by dividing material strengths by a safety factor λ_e that has to take account of all types of uncertainty, including those associated with the loading. In the now well-established "plastic" design method, the expected or "working" loads are multiplied by a load factor λ_p that has to take account of variability of materials, as well as of loads. Limit-state design puts these factors where they rightly belong, by using two sets of partial safety factors, γ_f for loads and γ_m for materials.

Design calculations consist essentially of checking that a stress or stress resultant due to a given load does not exceed the strength of a given material or structural member.❷ Fig. 1 illustrates the fact that the numerical values at which these comparisons are made are different for the three design philosophies. The dots represent numbers fed into the calculation; the arrows show the effect of λ_e、λ_p、γ_m and γ_f, on these numbers, and the crosses mark the points where the checking is done.

The specified or calculated loads and the specified or guaranteed strengths of materials are known as characteristic values. After multiplication by γ_f or division by γ_m, as the case may be, they become design values. There is as yet no uniform practice for setting out these

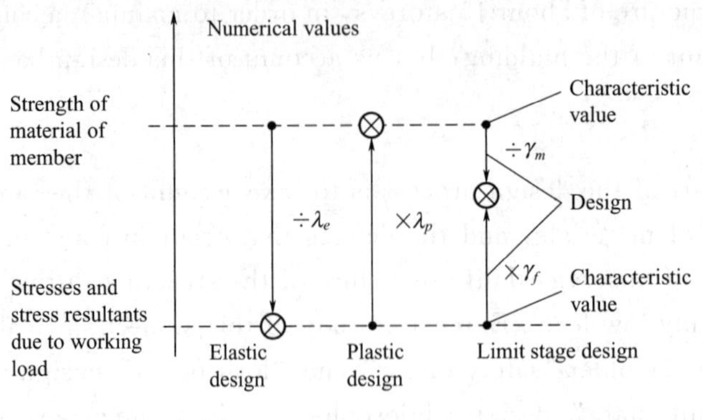

Fig. 1

calculations. In CP 110 the relevant values of γ_m are often incorporated in numerical coefficients, so that design equations include symbols such f_{cu} and f_y, which represent characteristic strengths of the concrete and reinforcement, respectively. The same practice has been followed for concrete in this book; but for steel, where γ_m values for certain types of member are not established, symbols are used for both characteristic and design strengths. The relationship are:

for reinforcement

$$f_{rd}\,(\text{design}) = f_{ry}\,(\text{yield}) \div \gamma_m$$

for rolled section

$$f_{sd}\,(\text{design}) = f_y\,(\text{yield}) \div \gamma_m$$

Limit-state design is also more rational than the two older methods in that it identifies more clearly the particular catastrophe or mode of unservice ability that each part of the design procedure is intended to avoid; and it provides a framework within which account can readily be taken of the different degrees of risk, uncertainty, or variability associated with various types of structure, loading, or materials.

These situations when something "goes wrong" are the limit states. They fall into two groups, which can be called "disasters" and "nuisances". Collapse, failure, overturning, buckling, or rupture of part or all of a structure is a disaster, which may lead to loss of life. These events are ultimatelimit states. The relevant partial safety factors all exceed 1.0, by amounts intended to ensure that the probability of such an occurrence is remote. Ultimate-strength and plastic methods of analysis are appropriate, since inelastic behaviour usually precedes failure; but elastic theory has to be used when no better method is available.

Excessive deflection or vibration, unsightly cracks, and spalling of concrete, due perhaps to corrosion of reinforcement, are nuisances, in that they may require repair or may limit the usefulness of the structure but they are not disasters. These are serviceability limit states. They normally occur while the structure is still elastic, so elastic analysis is appropriate. They must be avoided at working loads, so values of γ_f are usually 1.0. Their consequences are less severe than those of a failure, and to some extent (e.g., for deflections) they depend on average stiffnesses or strengths of materials, not on the occasional low value, so γ_m also can be taken as 1.0.

One situation that does not fit into this classification is fatigue failure under repeated loading. Initially a small fatigue crack is a form of unservice ability, but if allowed to spread it may eventually cause an ultimate limit state to be reached.❸

Limit-state design philosophy has been criticized on the ground that as the two limit states occur at different load levels, two sets of design calculations are needed, whereas with the older methods one was sufficient. This is only partly true, for it has been found possible when drafting codes of practice to identify many situations in which design for, say, the ultimate limit state will automatically ensure that certain types of unservice ability will not occur; and vice versa. In CP 110 the use of limiting span-depth ratios to control deflections and bar spacing to control cracking has made it unnecessary to analyze the structure at all for the serviceability loads.

Words and Phrases

1	deck	*n.*	甲板，舱面，桥面，层面
2	corrosion	*n.*	腐蚀，侵蚀
3	abrasion	*n.*	（表层）磨损处，磨损
4	monolithic	*a.*	（建筑等）庞大而无特点的，巨大而单调的
5	cladding	*n.*	镀层，保护层
6	reinforcement	*n.*	加强，强化
7	stanchion	*n.*	（用以支撑的）杆，支柱
8	buckling	*v.*	（使）搭扣扣住；（被）压垮，压弯
9	girder	*n.*	（桥梁或建筑物的）大梁，桁架
10	spalling	*n.*	剥落，脱落，散裂，剥离
11	deflection	*n.*	挠曲
12	nuisances	*n.*	麻烦事，妨害行为
13	premium	*n.*, *a.*	保险费，额外费用；高昂的，优质的
14	substantial	*a.*	大量的，价值巨大的，重大的，大而坚固的，结实的，牢固的
15	inflation	*n.*	通货膨胀，通胀率，充气
16	rupture	*n.*	破裂，断裂，爆裂
17	composite structures		组合结构

Notes

❶ Encasement in concrete is an economical method for steel columns, since the casing provides a substantial gain in strength.

对于钢柱来说，用混凝土包裹是一种经济的方法，因为套管使钢柱的强度大幅度提高。

❷ Design calculations consist essentially of checking that a stress or stress resultant due to a given load does not exceed the strength of a given material or structural member.

设计计算基本上包括检查由给定载荷产生的应力或应力的合力不会超过给定材料或结构构件的强度。

❸ One situation that does not fit into this classification is fatigue failure under repeated loading. Initially a small fatigue crack is a form of unservice ability, but if allowed to spread it may eventually cause an ultimate limit state to be reached.

有一种情况不符合这一分类，那就是反复加载下的疲劳破坏。在初始阶段，一个小的疲劳裂缝代表着结构失效的可能性，如果允许裂缝继续扩展，最终可能会引起结构达到极限状态。

Exercises

1. Translate the following into Chinese

(1) Excessive deflection or vibration, unsightly cracks, and spalling of concrete, due perhaps to corrosion of reinforcement, are nuisances, in that they may require repair or may limit the usefulness of the structure but they are not disasters.

(2) Limit-state design philosophy has been criticized on the ground that as the two limit states occur at different load levels, two sets of design calculations are needed, whereas with the older methods one was sufficient.

2. Translate the following into English

(1) 建筑和桥梁结构的设计主要是为了承担竖向荷载作用。

(2) 耐火程度是必须考虑的一个因素，因为它影响着设计者如何在混凝土结构、组合结构以及钢结构这些结构形式之间做出选择，而混凝土结构在这种条件下具有一定的优势。

Part Ⅲ The Reading Materials

Unit 1 Components of A Building

Materials and structural forms are combined to make up the various parts of a building, including the load-carrying frame, skin, floors, and partitions. The building also has mechanical and electrical systems, such as elevators, heating and cooling systems, and lighting systems. The superstructure is that part of a building above ground, and the substructure and foundation is that part of a building below ground.

The skyscraper owes its existence to two developments of the 19th century: steel skeleton construction and the passenger elevator. Steel is a construction material dates from the introduction of the Bessemer❶ converter in 1855. Gustame Eiffel (1832—1923) introduced steel construction in France. His designs for the Galerie des Machines and the Tower for the Paris Exposition of 1889 expressed the lightness of the steel framework. The Eiffel Tower, 984 feet (300 meters) high, was the tallest structure built by man and was not surpassed until 40 years later by a series of American skyscrapers.

The first elevator was installed by Elisha Otis in a department store in New York in 1857. In 1889, Eiffel installed the first elevators on a grand scale in the Eiffel Tower, whose hydraulic elevators could transport 2350 passengers to the summit every hour.

Load-carrying frame. Until the late 19th century, the exterior walls of a building were used as bearing walls to support the floors. This construction is essentially a post and lintel type, and it is still used in frame construction for houses. Bearing-wall construction limited the height of buildings because of the enormous wall thickness required;❷ for instance, the 16-story Monadnock Building built in the 1880's in Chicago had walls 5 feet (1.5 meters) thick at the lower floors. In 1883, William Le Baron Jenney (1832—1907) supported floors on cast-iron columns to form a cage-like construction. Skeleton construction, consisting of steel beams and columns, was first used in 1889. As a consequence of skeleton construction, the enclosing walls become a "curtain wall" rather than serving a supporting function. Masonry was the curtain wall material until the 1930's, when light metal and glass curtain walls were used. After the introduction of the steel skeleton, the height of buildings continued to increase rapidly.

All tall buildings were built with a skeleton of steel until World War Ⅱ. After the war, the shortage of steel and the improved quality of concrete led to tall buildings being built of reinforced concrete. Marina Towers (1962) in Chicago is the tallest concrete building in the United States; its height—588 feet (179 meters)—is exceeded by the 650 feet (198 meters) Post Office Tower in London and by other towers.

A change in attitude about skyscraper construction has brought a return to the use of the bearing wall. In New York city, the Columbia Broadcasting System Building, designed by

Eero Saarinen in 1962, has a perimeter wall consisting of 5 feet (1. 5 meter) wide concrete columns spaced 10 feet (3 meters) from column center to center. This perimeter wall, in effect, constitutes a bearing wall. One reason for this trend is that stiffness against the action of wind can be economically obtained by using the walls of the building as a tube; the World Trade Center buildings are another example of this tube approach. In contrast, rigid frames or vertical trusses are usually provided to give lateral stability.

Skin. The skin of a building consists of both transparent elements (windows) and opaque elements (walls). Windows are traditionally glass, although plastics are being used, especially in schools where breakage creates a maintenance problem. The wall elements, which are used to cover the structure and are supported by it, are built of a variety of materials: brick, precast, concrete, stone, opaque glass, plastics, steel, and aluminum. Wood is used mainly in house construction; it is not generally used for commercial, industrial, or public buildings because of the fire hazard.

Floors. The construction of the floors in a building depends on the basic structural frame that is used. In steel skeleton construction, floors are either slabs of concrete resting on steel beams or a deck consisting of corrugated steel with a concrete topping. In concrete construction, the floors are either slabs of concrete on concrete beams or a series of closely spaced concrete beams (ribs) in two directions topped with a thin concrete slab, giving the appearance of a waffle on its underside. The kind of floor that is used depends on the span between supporting columns or walls and function of the space. In an apartment building, for instance, where walls and columns are spaced at 12 to 18 feet (3. 7 to 5. 5 meters), the most popular construction is a solid concrete slab with no beams. The underside of the slab serves as the ceiling for the space below it. Corrugated steel decks are often used in office buildings because the corrugations, when enclosed by another sheet of metal, form ducts for telephone and electrical lines. ❸

Mechanical and Electrical Systems. A modern building not only contains the space for which it is intended (office, classroom, apartment) but also contains ancillary space for mechanical and electrical systems that help to provide a comfortable environment. These ancillary spaces in a skyscraper office building may constitute 25% of the total building area. The importance of heating, ventilating, electrical, and plumbing systems in an office building is shown by the fact that 40% of the construction budget is allocated to them. Because of the increased use of sealed buildings with windows that cannot be opened, elaborate mechanical systems are provided for ventilation and air conditioning. Ducts and pipes carry fresh air from central fan rooms and air conditioning machinery. The ceiling, which is suspended below the upper floor construction, conceals the ductwork and contains the lighting units. Electrical wiring for power and for telephone communication may also be located in this ceiling space or may be buried in the floor construction in pipes or conduits.

There have been attempts to incorporate the mechanical and electrical systems into the architecture of buildings by frankly expressing them; for example, the American Republic Insurance Company Building (1965) in Des Moines, Iowa, exposes both the ducts and the floor structure in an organized and elegant pattern and dispenses with the suspended ceiling. This type of approach makes it possible to reduce the cost of the building and permits innovations, such as in the span of the structure.

Soil and Foundations. All buildings are supported on the ground, and therefore the nature of the soil becomes an extremely important consideration in the design of any building. The design of a foundation depends on many soil factors, such as type of soil, soil stratification, thickness of soil layers and their compaction, and groundwater conditions. Soils rarely have a single composition; they generally are mixtures in layers of varying thickness. For evaluation, soils are graded according to particle size, which increases from silt to clay to sand to gravel to rock. In general, the larger particle soils will support heavier loads than the smaller ones. The hardest rock can support loads up to 100 tons per square foot (976.5 metric tons/sq meter), but the softest silt can support a load of only 0.25 ton per square foot (2.44 metric tons/sq. meter). All soils beneath the surface are in a state of compaction; that is, they are under a pressure that is equal to the weight of the soil column above it. ❹ Many soils (except for most sands and gavels) exhibit elastic properties—they deform when compressed under load and rebound when the load is removed. The elasticity of soils is often time-dependent, that is, deformations of the soil occur over a length of time, which may from minutes to years after a load is imposed. Over a period of time, a building may settle if it imposes a load on the soil greater than the natural compaction weight of the soil. Conversely, a building may heave if it imposes loads on the soil smaller than the natural compaction weight. The soil may also flow under the weight of a building; that is, it tends to be squeezed out.

Due to both the compaction and flow effects, buildings tend to settle. Uneven settlements, exemplified by the leaning towers in Pisa and Bologna, can have damaging effects—the building may lean, walls and partitions may crack, windows and doors may become inoperative, and, in the extreme, a building may collapse. Uniform settlements are not so serious, although extreme conditions, such as those in Mexico city, can have serious consequences. Over the past 100 years, a change in the groundwater level there has caused some buildings to settle more than 10 feet (3 meters). Because such movements can occur during and after construction, careful analysis of the soils under a building is vital.

The great variability of soils has led to a variety of solutions to the foundation problem. Where firm soil exists close to the surface, the simplest solution is to rest columns on a small slab of concrete (spread footing). Where the soil is softer, it is necessary to spread the column load over a greater area; in this case, a continuous slab of concrete (raft or mat) under the whole building is used. In cases where the soil near the surface is unable to support the weight of the building, piles of wood, or concrete are driven down to firm soil.

The construction of a building proceeds naturally from the foundation up to the superstructure. The design process, however, proceeds from the roof down to the foundation (in the direction of gravity). In the past, the foundation was not subjected to systematic investigation. A scientific approach to the design of foundations has been developed in the 20th century. Karl Terzaghi of the United States pioneered studies that made it possible to make accurate predictions of the behavior of foundations, using the science of soil mechanics coupled with exploration and testing procedures. Foundation failures of the past, such as the classical example of the leaning tower in Pisa, have become almost nonexistent. Foundations still are a hidden but costly part of many buildings.

Words and Phrases

1	partition	*n.*	分开，分割，隔墙，隔板
2	converter	*n.*	炼钢炉，吹风转炉
3	framework	*n.*	构架，框架，结构
4	surpass	*v.*	超过，胜过
5	exterior	*a.*	外部的，外面的，外部，表面
6	lintel	*n.*	楣，（门窗）过梁
7	opaque	*a.*	透明的，不透光的，不透明体
8	deck	*n.*	甲板，舱面，桥面，层面
9	corrugate	*v.*，*a.*	弄皱，使起皱纹；起皱的，起波纹的
10	duct	*n.*	管道，通道，预应力筋孔道
11	ancillary	*a.*	辅助的，附属的
12	ventilate	*v.*	使通风，使通气，给……装置通风设备
13	allocate	*v.*	分配，分派，配给
14	conceal	*v.*	隐蔽，隐瞒，把……隐藏起来
15	conduit	*n.*	管道，导管，水道，水管
16	stratification	*n.*	分层，层理
17	exhibit	*v.*	显示，呈现
18	curtain wall		悬墙，幕墙
19	bearing wall		承重墙
20	plumbing system		卫生设备系统
21	air conditioning		空气调节

Notes

❶ Bessemer：Henry Bessemer（1813—1898）英国冶金学家，1855 年提出不需外热的一种炼钢方法。这是首创大量生产钢的方法，对当时钢的生产起了一定的促进作用，其炼钢法叫做“贝色麦法”，又称“酸性底吹转炉炼钢法”。

❷ Bearing-wall construction limited the height of buildings because of the enormous wall thickness required.

因为所需墙体的厚度很大，承重墙结构限制了建筑物的高度。

❸ Corrugated steel decks are often used in office building because the corrugations, when enclosed by another sheet of metal, form ducts for telephone and electrical lines.

办公大楼中常使用波纹钢地板，这是因为波纹钢地板的波纹当有另一块金属板盖上时可以形成电话线和电线管道。

❹ All soils beneath the surface are in a state of compaction; that is, they are under a pressure that is equal to the weight of the soil column above it.

所有地表以下的土都处于受压状态，说得更精确些，这些土承受与作用在其上的土柱重量相等的压力。

Exercises

1. Translate the following into Chinese

(1) There have been attempts to incorporate the mechanical and electrical systems into the architecture of buildings by frankly expressing them; for example, the American Republic Insurance Company Building exposes both the ducts and the floor structure in an organized and elegant pattern and dispenses with the suspended ceiling.

(2) The wall elements, which are used to cover the structure and are supported by it, are built of a variety of materials: brick, precast, concrete, stone, opaque lass, plastics, steel, and aluminum. Wood is used mainly in house construction; it is not generally used for commercial, industrial, or public buildings because of the fire hazard.

2. Translate the following into English

(1) 由于框架结构的使用，封闭墙体更多地成为围护墙，而不是用作支撑构件。

(2) 在混凝土结构中，楼板或者是在混凝土梁上的混凝土板或者是在两个方向上有一组密跨的混凝土梁，上面有薄混凝土的板。

Unit 2 Building Materials

Materials for building must have certain physical properties to be structurally useful. Primarily, they must be able to carry a load, or weight, without changing shape permanently. When a load is applied to a structure member, it will deform; that is, a wire will stretch or a beam will bend. However, when the load is removed, the wire and the beam come back to the original positions. This material property is called elasticity. If a material were not elastic and a deformation were present in the structure after removal of the load, repeated loading and unloading eventually would increase the deformation to the point where the structure would become useless. All materials used in architectural structures such as stone and brick, wood, steel, aluminum, reinforced concrete, and plastics, behave elastically within a certain defined range of loading. If the loading is increased above the range, two types of behavior can occur: brittle and plastic. In the former the material will break suddenly. In the latter, the material begins to flow at a certain load (yield strength), ultimately leading to fracture. As examples, steel exhibits plastic behavior, and stone is brittle. The ultimate strength of a material is measured by the stress at which failure (fracture) occurs.

A second important property of a building material is its stiffness. This property is defined by the elastic modulus, which is the ratio of the stress (force per unit area), to the strain (deformation per unit length). The elastic modulus, therefore, is a measure of the resistance of a material to deformation under load. For two materials of equal area under the same load, the one with the higher elastic modulus has the smaller deformation. Structural steel, which has an elastic modulus of 30 million pounds per square inch (psi), or 2100000 kilograms per square centimeter, is 3 times as stiff as aluminum, 10 times as stiff as aluminum, 10 times as stiff as concrete, and 15 times as stiff as wood.

Masonry. Masonry consists of natural materials, such as stone, or manufactured products, such as brick and concrete blocks. Masonry has been used since ancient times; mud bricks were used in the city of Babylon for secular buildings, and stone was used for the great temples of the Nile Valley. The Great Pyramid in Egypt, standing 481 feet (147 meters) high, is the most spectacular masonry construction. Masonry units originally were stacked without using any bonding agent, but all modern masonry construction uses a cement mortar as a bonding material. Modern structural materials include stone, bricks of burnt clay or slate, and concrete blocks.

Masonry is essentially a compressive material; it cannot withstand a tensile force, that is, a pull. The ultimate compressive strength of bonded masonry depends on the strength of the masonry unit and the mortar. The ultimate strength will vary from 1000 to 4000 psi (70 to 280kg/sq. cm), depending on the particular combination of masonry unit and mortar used.

Timber. Timber is one of the earliest construction materials and one of the few natural materials with good tensile properties. Hundreds of different species of wood are found throughout the world, and each species exhibits different physical characteristics. Only a few species are used structurally as framing members in building construction. In the United States, for instance, out of more than 600 species of wood, only 20 species are used structurally. These are generally the conifers, or softwoods, both because of their abundance and because of the ease with which their wood can be shaped. The species of timber more commonly used in the United States for construction are Douglas fir, Southern pine, spruce, and redwood. The ultimate tensile strength of these species varies from 5000 to 8000 psi (350 to 560 kg/sq. cm). Hardwoods are used primarily for cabinetwork and for interior finishes such as floors.

Because of the cellular nature of wood, it is stronger along the grain than across the grain. Wood is particularly strong in tension and compression parallel to the grain, and it has great bending strength. These properties make it ideally suited for columns and beams in structures. Wood is not effectively used as a tensile member in a truss, however, because the tensile strength of a truss member depends upon connections between members. It is difficult to devise connections which do not depend on the shear or tearing strength along the grain, although numerous metal connectors have been produced to utilize the tensile strength of timbers. ❶

Steel. Steel is an outstanding structural material. It has a high strength on a pound-for-pound basis when compared to other materials, even though its volume-for-volume weight is more than ten times that of wood. It has a high elastic modulus, which results in small deformations under load. It can be formed by rolling into various structural shapes such as I-beams, plates, and sheets; it also can be cast into complex shapes; and it is also produced in the form of wire strands and ropes for use as cables in suspension bridges and suspended roofs, as elevator ropes, and as wires for prestressing concrete. Steel elements can be joined together by various means, such as bolting, riveting, or welding. Carbon steels are subject to corrosion through oxidation and must be protected from contact with the atmosphere by painting them or embedding them in concrete. Above temperatures of about 700℉ (371℃), steel rapidly loses its strength, and therefore it must be covered in a jacket of a fireproof material (usually concrete) to increase its fire resistance.

The addition of alloying elements, such as silicon or manganese, results in higher strength steels with tensile strengths up to 250000 psi (17500 kg/sq. cm). ❷ These steels are used where the size of a structural member becomes critical, as in the case of columns in a skyscraper.

Aluminum. Aluminum is especially useful as a building material when lightweight, strength, and corrosion resistance are all important factors. Because pure aluminum is extremely soft and ductile, alloying elements, such as magnesium, silicon, zinc, and copper, must be added to it to impart the strength required for structural use. Structural aluminum alloys behave elastically. They have an elastic modulus one third as great as steel and therefore deform three times as much as steel under the same load. The unit weight of an aluminum alloy is one third that of steel, and therefore an aluminum member will be lighter than a steel member of comparable strength. The ultimate tensile strength of aluminum alloys ranges from 20000 to 60000 psi (1400 to 4200 kg/sq. cm).

Aluminum can be formed into a variety of shapes; it can be extruded to form I-beams, drawn to form wire and rods, and rolled to form foil and plates. Aluminum members can be put together in the same way as steel by riveting, bolting, and (to a lesser extent) by welding. Apart from its use for framing members in buildings and prefabricated housing, aluminum also finds extensive use for window frames and for the skin of the building in curtain-wall construction.

Concrete. Concrete is a mixture of water, sand and gravel, and Portland cement. Crushed stone, manufactured lightweight stone, and seashells are often used in lieu of natural gravel. Portland cement, which is a mixture of materials containing calcium and clay, is heated in a kiln and then pulverized. Concrete derives its strength from the fact that pulverized Portland cement, when mixed with water, hardens by a process called hydration. In an ideal mixture, concrete consists of about three fourths sand and gravel (aggregate) by volume and one fourth cement paste. The physical properties of concrete are highly sensitive to variations in the mixture of the components, so a particular combination of these ingredients must be custom-designed to achieve specified results in terms of strength or shrinkage. When concrete is poured into a mold or form, it contains free water, not required for hydration, which evaporates. As the concrete hardens, it releases this excess water over a period of time and shrinks. As a result of this shrinkage, fine cracks often develop. In order to minimize these shrinkage cracks, concrete must be hardened by keeping it moist for at least 5 days. The strength of concrete increases in time because the hydration process continues for years; as a practical matter, the strength at 28 days is considered standard.

Concrete deforms under load in an elastic manner. Although its elastic modulus is one tenth that of steel, similar deformations will result since its strength is also about one tenth that of steel. Concrete is basically a compressive material and has negligible tensile strength.

Reinforced concrete. Reinforced concrete has steel bars that are placed in a concrete member to carry tensile forces. These reinforced bars, which range in diameter from 0.25 inch (0.64cm) to 2.25 inches (5.7cm), have wrinkles on the surfaces to ensure a bond with the concrete. Although reinforced concrete was developed in many countries, its discovery usually is attributed to Jseph Monnier, a French gardener, who used a wire network to reinforce concrete tubes in 1868. This process is workable because steel and concrete expand and contract equally when the temperature changes. If this were not the case, the bond between the steel and concrete would be broken by a change in temperature since the two materials would respond differently. Reinforced concrete can be molded into innumerable shapes, such as beams, columns slab, and arches, and is therefore easily adapted to a particular form of building. ❸ Reinforced concrete with ultimate tensile strength in excess of 10000 psi (700 kg/sq. cm) is possible, although most commercial concrete is produced with strength under 6000 psi (420 kg/sq. cm).

Plastics. Plastics are rapidly becoming important construction materials because of great variety, strength, durability, and lightness. A plastic is a synthetic material or which can be molded into any desired shape and which uses an organic substance as a binder. Organic plastics are divided into two general groups: thermosetting and thermoplastic. The thermosetting group becomes rigid through a chemical change that occurs when heat is applied; once set, these plastics cannot be remolded. The thermoplastic group remains soft at high temperatures and must be cooled before becoming rigid; this group is not used generally as a structural material. The ultimate strength of most plastic materials is from 7000 to 12000 psi (490 to 840 kg/sq. cm), although nylon has a tensile strength up to 60000 psi (4200kg/sq. cm).

Words and Phrases

1	elasticity	*n.*	弹性，弹力，弹性力学
2	stiffness	*n.*	劲度，刚度，硬度
3	secular	*a.*	世俗的，现世的，非宗教的
4	temple	*n.*	庙，寺，神殿，教堂
5	pyramid	*n.*	金字塔，四面体
6	stack	*v.*	堆叠，成堆，整齐地堆起
7	slate	*n.*	板岩，石板瓦，铺石板
8	timber	*n.*	木材，木料，原木
9	conifer	*n.*	针叶树，松柏类植物
10	abundance	*a.*	丰富，充裕，大量
11	spruce	*n.*	云杉，云杉木
12	cabinetwork	*n.*	细木工，细木家具
13	cellular	*a.*	细胞的，分格的，多孔状的
14	grain	*n.*	颗粒，纹理，粒面
15	silicon	*n.*	硅，硅元素
16	manganese	*n.*	（化学元素）锰
17	seashell	*n.*	海贝，贝壳
18	negligible	*a.*	可以忽略的，微不足道的
19	synthetic	*a.*	合成的，人造的，人造树脂
20	resin	*a.*	树脂，胶质，人造树脂
21	thermosetting	*a.*	热凝性的，热固性的
22	elastic modulus	*n.*	弹性模量
23	Douglas fir	*n.*	美国松
24	ultimate tensile strength		极限抗拉强度

Notes

❶ It is difficult to devise connections which do not depend on the shear or tearing strength along the grain, although numerous metal connectors have been produced to utilize the tensile strength of timbers.

虽然生产了许多利用木材抗拉强度的金属连接器，但很难设计与沿木材顺纹方向抗剪强度或抗扯裂强度关系不大的接头。

❷ The addition of alloying elements, such as silicon or manganese, results in higher strength steels with tensile strengths up to 250000 psi.

添加像硅或锰这样的合金元素，会得到抗拉强度达 250000 磅/平方英寸的高强钢筋。

❸ Reinforced concrete can be molded into innumerable shapes, such as beams, columns, slabs, and arches, and is therefore easily adapted to a particular form of building.

钢筋混凝土可被浇铸成各种形状，例如梁、柱、板和拱，因此它宜适用于建筑的特殊结构。

Exercises

1. Translate the following into Chinese

(1) When a load is applied to a structure member, it will deform; that is, a wire will stretch or a beam will bend. However, when the load is removed, the wire and the beam come back to the original positions. This material property is called elasticity.

(2) The elastic modulus is a measure of the resistance of a material to deformation under load. For two materials of equal area under the same load, the one with the higher elastic modulus has the smaller deformation.

2. Translate the following into English

(1) 材料的极限强度是通过破坏（断裂）发生位置的应力来测定的。

(2) 木材是最早的结构材料，也是少有的几种具有较好受拉性能的天然材料之一。

Unit 3 Special Concrete

Lightweight Concrete

Structural lightweight concrete is structural concrete in every respect except that for reasons of overall cost economy. It is made with light- weight cellular aggregates so that its unit weight is approximately two-thirds of the unit weight of concrete made with typical natural aggregates. Since light weight, and not strength, is the primary objective, the specifications limit the maximum permissible unit weight of concrete. Also, sine highly porous aggregates tend to reduce concrete strength greatly 5 the specifications require a minimum 28-day compressive strength to ensure that the concrete is of structural quality.

ACI 213R-79, Guide for Structural Lightweight Aggregate Concrete, defines structural lightweight aggregate concretes as concretes having a 28-day compressive strength in excess of 2500 psi (17 MPa) and a 28-day, air dried, unit weight not exceeding 115 lb/ft (1850 kg/m). The concrete may consist entirely of lightweight aggregates or, for various reasons, a combination of lightweight and normal-weight aggregates. From the standpoint of workability and other properties, it is a common practice to use normal sand as fine aggregate, and to limit the nominal size of the lightweight coarse aggregate to a maximum of 3/4 in. According to ASTMC 330, fine lightweight and coarse lightweight aggregates are required to have dry-loose weights not exceeding 70 and 55 lb/ft., respectively. The specification also contains requirements with respect to grading, deleterious substances, and concrete-making properties of aggregate, such as strength, unit weight, drying shrinkage, and durability of concrete containing the aggregate.

The basic economy of lightweight concrete can be demonstrated by the savings in reinforcing. In ordinary reinforced concrete the eco-nomic advantage is not as pronounced as in prestressed concrete. The prestressing force in most cases is computed strictly from the dead load of the structure; consequently, a weight reduction of 25 percent results in a substantial reduction in the weight of prestressing tendons. Among other advantages of reduction in weight of concrete is the superior resistance of shear elements to earthquake loading since seismic forces are largely a direct function of the dead weight of a structure.

High-strength Concrete

For mixtures made with normal-weight aggregates, high-strength concretes are considered to be those which have compressive strengths in excess of 6000 psi (40 MPa). Two arguments are advanced to justify this definition of high-strength concrete.

(1) The bulk of the conventional concrete is in the range 3000 to 6000 psi. To produce concrete above 6000 psi, more stringent quality control and more care in the selection and proportioning of materials (plasticizers, mineral admixtures, type and size of aggregates,

etc.), are needed. Thus, to distinguish this specially formulated concrete, which has a compressive strength above 6000 psi, it should be called high strength.

(2) Experimental studies show that in many respects the microstructure and properties of concrete with a compressive strength above 6000 psi are considerably different from those of conventional concrete. Since the latter is the basis of current concrete design practice (e. g., the empirical equation for estimating the elastic modulus from compressive strength), the designer will be alerted if concrete of higher than 6000 psi is treated as a separate class.

The use of the highest-possible strength concrete and minimum steel offers the most economical solution for columns of high-rise buildings. The economic advantage gained by the use of high-strength concrete for columns and shear walls was clearly demonstrated by several structures built in Chicago, New York, Houston, and other cities in the United States. In the United States during the past 20 years, high-strength concrete has been used mainly for constructing the reinforced concrete frames of buildings 30 stories and higher. In the precast and prestressed concrete industries, the use of high-strength concrete has resulted in a rapid turnover of molds, higher productivity, and less loss of products during handling and transportation. Since their permeability is very low, high-strength concretes also find application where durability of concrete is adversely affected due to abrasion, erosion, or various chemical attacks.

High-workability Concrete

For want of a standard definition, high-workability concrete may be considered as the concrete of a flowing consistency (7 to 9 in. of slump), which can be placed and compacted with little or no effort and which is, at the same time, cohesive enough to be handled without segregation and bleeding. Architects and engineers who have worked with superplasticized concretes to achieve high strength or high workability believe that the advent of superplasticizers has started a new era in the construction practice by extending the use of concrete to structures with complex designs and highly demanding applications. Superplasticizing chemicals are expensive; their use may increase the materials cost of a concrete mixture, but the increased cost is easily offset by savings in labor cost that are made possible by maximum utilization of on-site time. The benefits from the use of flowing concrete are as follows:

(1) Placing of concrete with reduced vibration in areas of closely bunched reinforcement and in areas of poor access (by using flowing concrete, the need to cut or adopt formwork to obtain vibrator access may be obviated).

(2) The capability of placing-very rapidly, easily, and without vibration concrete for bay areas, floor slabs, roof decks, and similar structures.

(3) The very rapid pumping of concrete.

(4) Placing concrete by means of tremie pipe.

(5) The production of uniform and compact concrete surfaces.

...

Unit 4 The Procedures of Structural Design

Although each design problem is different and has its own specific peculiarities, the solution passes through several sequential steps or phases which are similar in nature from problem to problem. To give the student or young engineer a better idea of what is involved in a design these steps are briefly discussed below.

Performance Requirements

The designer must make a thorough study of the technological and service performance requirements that must be expected from the structure. Matters such as load intensities and their duration, any dynamic actions that might take place, clearance requirements if any, and loading combinations are some of these requirements.

Local Conditions

The design of any civil engineering structure is greatly influenced by local conditions. Even if the performance requirements are the same and the same type of structure is selected for different locations, each structure will exhibit differences in design so as to comply with the local conditions. Site elevations, environment, foundation characteristics, water tables, immediate surroundings, and the environment as a whole represent some of the local conditions that need to be investigated.

By considering the local conditions and the performance requirements of the structure, the designer will arrive at different possible layouts, structural systems, and their geometry. Sketches combined with some preliminary sizing based on experience and rough calculations enable him to select a structure. This creative process based on past experience, and a sound knowledge of the behavior of different structural systems under the prescribed performance requirements is difficult to standardize. It takes into account aesthetics, economy, local labor and material conditions, structural behavior under expected loadings, and possible dynamic effects, to name but a few factors. No hard and fast rules can be laid down. The ability to recognize the various factors that may affect a given design is based not only on the gathering of the necessary technical information, but also on ingenuity, intuition, and above all experience.

Preliminary Design

Preliminary sizes, or ratios of sizes, have to be assumed before any analysis can take place. Size ratios depend on the type of structure, its redundancy, and the major design procedure that is adopted. Sizes refer to cross-sectional areas, or to moments of inertia for use in the allowable stress design method and ultimate load capacities for use in the plastic design method. Again, experience and comparison with similar designs and use of available empirical rules combined with some rough calculations are the main tools used to arrive at a preliminary design.

Structural Analysis

The forces and moments to be expected in the preliminary design can now be determined. In most cases strength is the initial design criterion, while all other criteria are used to carry out checks at a later stage.

Selection of Shapes and Sizes

On the basis of the analysis thus performed a revised selection of sizes and shapes and their ratios takes place. These are compared with those initially assumed. To arrive at an economical design it is usually necessary now to repeat the analysis with the revised sizes and shapes. This trial-and-error method of assuming dimensions and checking the resulting stress by means of an analysis is typical of any design. Much busy work can be avoided by making use of past experience for the initial sizing; also, computer programs can be used to carry out this trial-and-error procedure to arrive at an optimum design.

Secondary Design Consideration

Most designs are initially based on the strength criterion (first yield or structural collapse). Local stability considerations usually are automatically included in this criterion. Overall structural stability, deformations, dynamic behavior, the occurrence of secondary stresses due to temperature changes, and the design and detailing of connections could all be classified as secondary design criteria. Upon investigating these so-called secondary effects, they prove to be major effects requiring modifications or drastic revisions of the initial strength design. Based on experience the engineer may be able to recognize beforehand any of such governing criteria and develop an initial design from these rather than on the basis of strength, which now instead is treated as a secondary criterion.

Unit 5 Reinforced Concrete Columns in a Frame

The design or analysis of a reinforced concrete column must be undertaken in conjunction with that of the entire structure. In general the same simplifying assumptions must therefore be used. These may range from using the same EI for all members in a frame, ignoring size and reinforcing variations, to only smearing composite nominal tangent EIs over short segments of each member, while taking into account variations in cracking and load levels, as well as the effects of shrinkage, creep, reinforcement distribution and the nonlinear short time stress-strain responses of the materials.

Using any desired assumptions, it is necessary to obtain a member stiffness matrix in order to find the axial load and moments and shears which the member will have to resist prior to collapse at ultimate load, and with deformations and crack widths less than their respective permissible values at service loads. Member cross sections designed to accomplish this can then be used to obtain a new stiffness matrix for use in a new cycle of analysis and design. The shear strength, shown to be a factor of utmost importance in earthquake resistance; is traditionally dealt with separately. The fact that there is biaxial rather than uniaxial bending in the column need not change the shear analysis. The member must be analyzed for the resultant shear force.

Given the previously stated requirements, it is obvious that the mere design of the column cross section for strength under some combination of "ultimate" loads is quite insufficient. However, the strength and behavior of the cross section represent anchor points in the analysis and design of reinforced concrete columns. In this chapter a variety of methods for the design and analysis of slender and nonslender reinforced concrete columns will be presented and examined, some simple and inaccurate, others more complex and, hopefully, more accurate.

Given the nature of the problem, a combination of simplicity and accuracy is probably impossible to achieve, though it appears that satisfactory performance of columns can be obtained quite economically. The design and analysis of reinforced concrete columns under combined axial load and uniaxial bending is discussed in many textbooks on reinforced concrete structures (Wang and salmon, 1985). Biaxial bending introduces complications in inverse ratio to the number of simplifying assumptions to be made in the structural analysis. If the simplest method is used, in which fully elastic, uncracked, short column behavior is assumed, neither the presence of axial load nor that of bending in the orthogonal direction will affect the stiffness of the member in one direction, and cross section stresses may be obtained from uncoupled bidirectional frame analyses and superposed. Many columns designed in this way are serving successfully all over the world.

At the other end of the spectrum of complexities, the length of the column is divided into short segments and a grid is superimposed on the cross section of each. The time and stress dependent responses of both the concrete and the individual reinforcing bars are tracked in a historical loading sequence, and axial load dependent, biaxial moment-curvature relationships of each segment are calculated. Numerical integration over the member length is then used to obtain a member stiffness matrix for just that history and set of loading conditions. Enormous computer resources are required to iterate through an entire structural frame in this manner and no real building columns have yet been designed using it. Very small frames have, however, been analyzed by Gesund (1967) and Shah and Gesund (1972).

Between the two extremes, a number of useful methods of design, analysis, redesign, reanalysis, etc., have been developed. Several have proven to yield safe and economical designs in practice. Almost all Codes of Practice require design for "ultimate" loads, but recommend that an elastic method of frame analysis be used. This defies logic, since the structure cross sections will be anything but elastic at the loads for which the analysis is being carried out. They will, in fact, have been designed for large yielding of the reinforcement and incipient crushing of the compression concrete, i. e. considerable hinge rotations. Nevertheless, for frame members not affected by slenderness and/or sidesway deformations, the method has been found to be workable though, lacking controls on hinge rotations, not always conservative, and it has permitted the economical design of a large number of serviceable structures. It is doubtful that it should be used with the same confidence when story drifts and member or frame stability become considerable problems.

Unit 6 Structural Reliability

Structural reliability theory is concerned with the rational treatment of uncertainties in structural engineering and with the methods for assessing the safety and serviceability of civil engineering and other structures. It is a subject which has grown rapidly during the last decade and has evolved from being a topic for academic research to a set of well-developed or developing methodologies with a wide range of practical applications.

Uncertainties exist in most areas of civil and structural engineering and rational design decisions cannot be made without modeling them and taking them into account. Many structural engineers are shielded from having to think about such problems at least when designing simple structures, because of the prescriptive and essentially deterministic nature of most codes of practice. This is an undesirable situation. Most loads and other structural design parameters are rarely known with certainty and should be regarded as random variables or stochastic processes, even if in design calculations they are eventually treated as deterministic. Some problems such as the analysis of load combination cannot even be formulated without recourse to probabilistic reasoning.

Until fairly recently there has been a tendency for structural engineering to be dominated by deterministic thinking, characterized in design calculations by the use of specified minimum material properties, specified load intensities and by prescribed procedures for computing stresses and deflections. This deterministic approach has almost certainly been reinforced by the very large extent to which structural engineering design is codified and the lack of feedback about the actual performance of structures. For example, actual stresses are rarely known, deflections are rarely observed or monitored, and since most structures do not collapse the real reserves of strengths are generally not known. In contrast, in the field of hydraulic systems, much more is known about the actual performance of, say, pipe networks, weirs, spillways etc., as their performance in service can be relatively easily observed or determined.

Most structural design is undertaken in accordance with codes of practice, which in many countries have legal status, meaning that compliance with the code automatically ensures compliance with the relevant clauses of the building laws. Structural codes typically and properly have a deterministic format and describe what are considered to be the minimum standards for design, construction and workmanship for each type of structure. Most codes can be seen to be evolutionary in nature 9 with changes being introduced or major revisions made at intervals of 3-10 years to allow for: new types of structural form, the effects of improved understanding of structural behaviour, the effects of changes in manufacturing tolerances or quality control procedures, a better knowledge of loads, etc. The lack of information about the actual behaviour of structures combined with the use of codes embodying relatively high safety factors can lead to the view, still held by some engineers as well as by some members of the general public, that absolute safety can be achieved. Absolute safety is of course unobtainable; and such a goal is also undesirable since absolute safety could be achieved only by deploying infinite resources. It is now widely recognized, however, that some risk of unacceptable structural performance must be tolerated. The main object of structural design is therefore to ensure, at an acceptable level of probability, that each structure will not become unfit for its intended purpose at any time during its specified design life. Most structures however, have multiple performance requirements, commonly expressed in terms of a set of serviceability and ultimate limit states, most of which are not independent; and thus the problem is much more complex than the specification of just a single probability.

There is a need for all structural engineers to develop an understanding of structural reliability theory and for this to be applied in design and construction, either indirectly through codes or by direct application in the case of special structures having large failure consequences, the aim in both cases being to achieve economy together with an appropriate degree of safety. The subject is now sufficiently well developed for it to be included as a formal part of the training of all civil and structural engineers, both at undergraduate and post-graduate levels. Courses on structural safety have been given at some universities for a number of years.

Unit 7 Bond of Prestressing Tendons

Two types of bond stress must be considered in the case of prestressed concrete. The first of these is referred to as "transfer bond stress" and has the function of transferring the force in a pre-tensioned tendon to the concrete. Transfer bond stresses come into existence when the prestressing force in the tendons is transferred from the prestressing beds to the concrete section. The second type of bond is termed "flexural bond stress" and comes into existence in pre-tensioned and bonded, post-tensioned members when the members are subjected to external loads. Flexural bond stress does not exist in unbonded, post-tensioned construction, which accounts for the term "unbonded". When a prestressing tendon is stressed, the elongation of the tendon is accompanied by a reduction in the diameter due to Poisson's effect.

When the tendon is released, the diameter increases to its original diameter at the ends of the prestressed member where it is not restrained. This phenomenon is generally regarded as a primary factor that influences the bonding of pre-tensioned wires to the concrete. The stress in the wire is zero at the extreme end and is at a maximum value at some distance from the end of the member.

Therefore, in the length of the tendon from the extreme end to the point where it attains maximum stress, called the "transmission length", there is a gradual decrease in the diameter of the tendon, giving the tendon a slight wedge shape over this length. This wedge shape is often referred to as the "Hoyer Effect" after the German engineer E. Hoyer, who was one of the early engineers to develop this theory. Hoyer, and others more recently, derived elastic theory to compute the transmission length as a function of Poisson's ratio for steel and concrete, the moduli of elasticity of steel and concrete, the diam-eter of the tendon, the coefficient of friction between the ten-don and the concrete, and the initial and effective stresses in the steel. Laboratory studies of the transmission lengths have indicated a relatively close agreement between the theoretical and actual values. There can be wide variation, however, due to the different properties of concrete and steel and due to surface conditions of the tendons, which affect the coefficient of friction.

There is reason to believe that the configuration of a seven-wire strand (i. e., 6 small wires twisted about a slightly larger center wire) results in very good bond characteristics. It is believed the Hoyer Effect is partially responsible for this, but the relatively large surface area and twisted configuration is believed to result in a significant mechanical bond.

Although these theoretical relationships are of academic interest, they have little practical application, due to the inability of designers and fabricators of prestressed concrete to control the several factors that influence the transmission length. Fortunately, there has been sufficient research into the magnitude of transmission lengths under both laboratory and production conditions for the following significant conclusions to be drawn.

(1) The bond of clean three and seven-wire prestressing strands and concrete is adequate for the majority of pre-tensioned concrete elements.

(2) Members that are of such a nature that high moments may occur near the ends of the members, such as short cantilevers, require special consideration.

(3) Clean smooth wires of small diameter are also adequate for use in pretensioning, but the transmission length for tendons of this type should be expected to be approximately double that for seven-wire strands (expressed as a multiple of the diameter).

(4) Under normal conditions, the transmission length for clean seven-wire strands can be assumed to be equal to 50 times the diameter of the strand.

(5) The transmission length of tendons can be expected to increase from 5 to 200.6 within one year after release as a result of relaxation.

(6) The transmission length of tendons released by flame cutting or with an abrasive wheel can be expected to be from 20% to 30% greater than tendons that are released gradually.

(7) Hard non-flaky surface rust and surface indentations effectively reduce the transmission lengths required for strand and some forms of wire tendons.

(8) Concrete-compressive strengths between 1500 and 5000 psi at the time of release result in transmission lengths of the same order, except for strand tendons larger than 1/2 in., in which, case strengths less than 3000 psi result in larger transmission lengths.

(9) It would seem prudent to use 3000 psi as a minimum release strength in pre-tensioned tendons, except for very unusual cases. Higher strengths may be required for tendons larger than 1/2 in.

(10) Because of relaxation, a small length of tendon (3″1) at the end of a member can be expected to become completely unstressed.

(11) The degree of compaction of the concrete at the ends of pre-tensioned members is extremely important if good bond and short transmission lengths are to be obtained. Honeycombing must be avoided at the ends of the beams.

(12) There is little if any reason to believe that the use of end blocks improves the transfer bond of pre-tensioned tendons, other than to facilitate the placing and compacting of the concrete at the ends. Hence, the use of end blocks is considered unnecessary in pre-tensioned beams, if sufficient care is given to this consideration.

(13) Tensile stresses and strains develop in the ends of pre-tensioned members along the transmission length as a result of the wedge effect of the tendons. Little if any beneficial results can be gained in attempting to reduce these stresses and strains by providing mild reinforcing steel around the ends of the tendons, since the concrete must undergo large deformations and would probably crack before such reinforcing steel could become effective.

(14) Lubricants and dirt on the surface of tendons has a detrimental effect on the bond characteristics of the tendons.

Bond stresses also occur between the tendons and the concrete in both pretensioned and bonded, post-tensioned members, as a result of changes in the external load. There are of course no transfer bond stresses in post-tensioned members, since the end anchorage device accomplishes the transfer of stress. Although it is known that flexural-bond stresses are relatively low in prestressed members for loads less than the cracking load, there is an abrupt and significant increase in these bond stresses after the cracking load is exceeded. Because of the indeterminancy which results from the plasticity of the concrete for loads exceeding the cracking load, accurate computation of the flexural-bond stresses cannot be made under such conditions. Again, testes must be relied upon as a guide for engineering design; however, the earthquake intensity rather than the magnitude is used. The intensity is a measure of earthquake's destructiveness. The intensity depends primarily on the magnitude and the distance between the epicenter and the location where the intensity is evaluated. The common measure for the intensity is the Modified Mercalli Intensity Scale. However, in engineering design, the ground acceleration is often used as the intensity and applied to the structure as the base excition intensity level. Empirical relations, known as attenuation equations, relate the intensity at a desired location to the magnitude and distance.

Unit 8　Cable Structure

Introduction

The use of materials in both tensile and compressive structures is extremely efficient since the entire cross section of each member is uniformly stressed. The strength of the members in a tensile structure is limited only by the basic material strength; thus, these are generally lightweight systems that are economical for spanning large distances. Since the members in compressive structures have a propensity for buckling, the material stress is limited to a level which is lower than that for tensile members. The arch is a common type of compressive structure, but under some conditions the internal state of stress can include shear and bending moment. Nevertheless, compression is usually the dominant internal force.

Cables with Concentrated Forces

A lightweight flexible cable loaded with only a single concentrated force will assume the configuration of the equilibrium force polygon. The term funicular, from the Latin words for chord, is used to describe this situation. Thus, a funicular polygon is the shape assumed by a string supporting concentrated forces, and a funicular structure is one in which the loads are transmitted by either tension or compression. Funicular construction is frequently used to describe the graphical approach for obtaining the bending-moment diagram for a loaded beam.

Consider the cable in Fig. 1; the ends are supported at different vertical elevations, and it id loaded with a single concentrated vertical force. Assume that the weight of the cable is negligible and that it is flexible enough to offer no bending resistance. The horizontal component of tension in the cable at any point is constant since only a vertical load is applied; therefore,

$$R_{ax}=R_{bx}=H \tag{1}$$

Summing moments about point a for the total cable gives

$$LR_{by}=P_i x_i+H(y_b-y_a) \tag{2}$$

where y_a and y_b are the y coordinates of points a and b, respectively, and x_i is the horizontal distance to the applied force P_i. Examining a typical point on the cable at point x, where $x \leqslant x_i$, and summing moments about this point using the free-body diagram to the right yields.

$$R_{by}(L-x)=H(y_b-y)+P_i(x_i-x) \tag{3}$$

where L is the horizontal distance between the supports and y is the vertical coordinate of the generic point, substituting Eq. (2) into Eq. (3) and rearranging terms gives

$$H[y_a+(y_b-y_a)x/L-y]=P_i x(1-x_i/L) \tag{4}$$

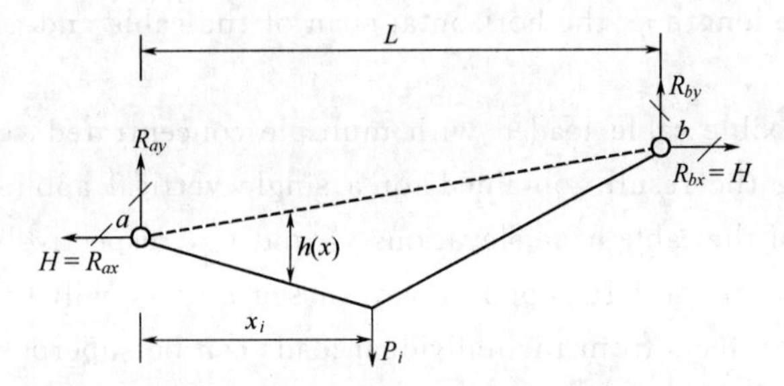

Fig. 1 A cable with a single concentrated vertical load

Note that the terms in the square brackets on the left-hand side represent the vertical distance from the point on the cable to the line joining the two support points. This is designated $h(x)$ and referred to as the sag of the cable. Substituting $h(x)$ into Eq. (4) gives the simplified form

$$Hh(x)=p_i x(1-x_i/L) \qquad \text{for } x \leqslant x_i \tag{5}$$

A similar expression for a point to the right of x_i can be obtained in the same manner. Thus, examining a point for which $x \leqslant x_i$, taking moments about the point using the free-body diagram to the right, substituting R_{by} from Eq. (2) into the resulting equation, rearranging terms, and using the cable sag $h(x)$, we have

$$Hh(x)=p_i x_i(1-x_i/L) \qquad \text{for } x \geqslant x_i \tag{6}$$

Eqs. (5) and Eqs. (6) can be interpreted by comparing them with the response of a simply supported beam of length L acted upon by the same loading as the cable (Fig. 2). The moment diagram for this beam is the illustrated bilinear shape, with

$$M(x)=p_i x(1-x_i/L) \qquad \text{for } x \leqslant x_i \tag{7}$$

$$M(x)=p_i x_i(1-x_i/L) \qquad \text{for } x \geqslant x_i \tag{8}$$

A comparison of Eqs. (5) and (6) and Eqs. (7) and (8), respectively, makes the following statement possible.

If a cable with negligible weight and transverse stiffness is acted on by a single vertical concentrated force, the product of the horizontal component of cable tension and the sag at any point is equal to the bending moment at a corresponding section of a simply supported

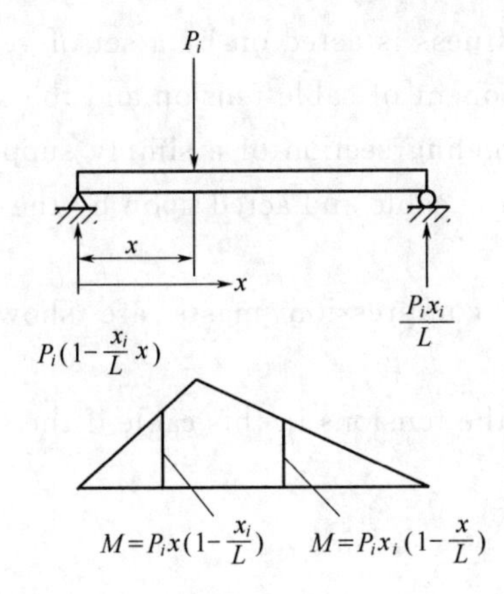

Fig. 2 Beam loaded with a single concentrated load and the associated moment diagram

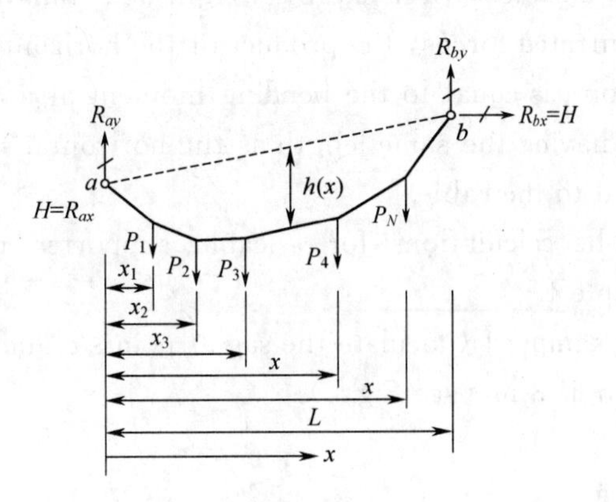

Fig. 3 A cable with multiple concentrated vertical loads

beam having the same length as the horizontal span of the cable and acted upon by the load applied to the cable.

A lightweight flexible cable loaded with multiple concentrated vertical forces (Fig. 3) can be analyzed using the results obtained for a single vertical applied load. The left and right support points of the cable have elevations y_a and y_b, respectively, and a generic point located a distance x from the left support, with a sag $h(x)$, will be examined. By using Eqs. (5) and (6) the effects from the individual loads can be superposed to obtain the total response of this cable. Thus, for the loads to the left of $x(x \geqslant x_i)$, Eq. (6) can be used for each force and the results added to give

$$[Hh(x)]_L = \left(1 - \frac{x}{L}\right) \sum_{i=1}^{N_L} P_i x_i \tag{9}$$

where $[Hh(x)]_L$ is the total effect of the N_L forces to the left of the point, for example, $N_L = 3$ for the cable in Fig. 3 Similarly, for the N_R forces to the right of $x(x \geqslant x_i)$ their effect$[Hh(x)]_L$ is obtained using Eq. (5), which yields

$$[Hh(x)]_R = \frac{x}{L} \sum_{j=1}^{N_R} P_j (L - x_i) \tag{10}$$

Thus, Noting that the total number of vertical forces $N = N_L + N_R$ and simplifying, the product of the total horizontal component of the cable tension H and the sag h is

$$\begin{aligned} Hh(x) &= [Hh(x)]_L + [Hh(x)]_R \\ &= \left(1 - \frac{x}{L}\right) \sum_1^N M_a - \sum_1^{N_R} M_R \end{aligned} \tag{11}$$

Where, $\sum_1^N M_a$ represents the moment of all the applied forces (P_1, P_2, …, P_b, …, P_N) taken about point a and $\sum_1^{N_R} M_R$ is the moment of all the forces to the right of point x about point x ($P_4, P_5, \cdots, P_b, \cdots, P_N$ for the cable in Fig. 3). The right-hand side of Eq. (11) can be interpreted as the moment at a generic cross section of a simply supported beam with a span L and N applied loads.

If a cable with negligible weight and transverse stiffness is acted on by a set of vertical concentrated forces, the product of the horizontal component of cable tension and the sag at any point is equal to the bending moment at a corresponding section of a simply supported beam having the same length as the horizontal span of the cable and acted upon by the loads applied to the cable.

The calculations for a cable supported on two compression masts are shown in Example 1.

Example 1 Calculate the sag at points c and d and the tensions in this cable if the sag at point d is 3 m (see Fig. 4).

Solution

The bending moment in. the beam equals the product of H and the sag, thus at point d.

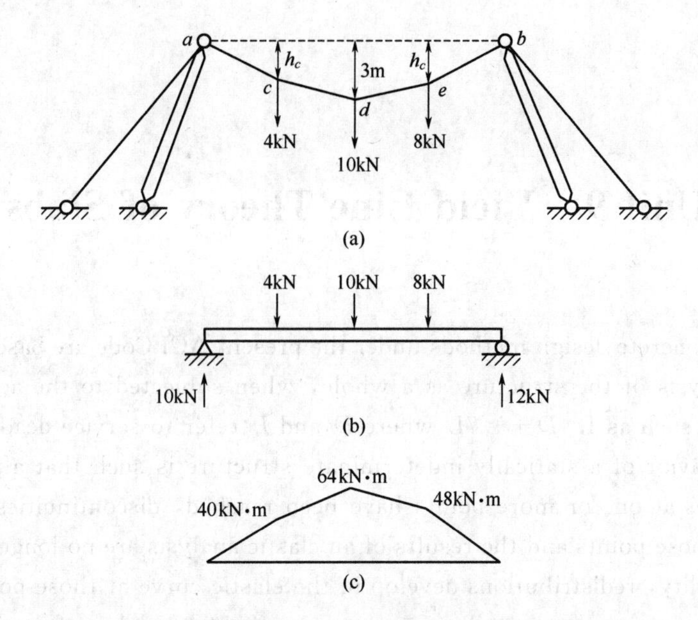

Fig. 4 The analysis of cable structure with vertical loads

(a) Cable structure; (b) Corresponding beam; (c) Beam moment diagram

$$3H=64 \qquad H=21.3333\text{kN}$$

At point c

$$21.333h_c=40 \qquad h_c=1.875\text{m}$$

At point e

$$21.333h_e=48 \qquad h_e=2.250\text{m}$$

The cable tensions are calculated by. dividing H by the cosine of the angle between the cable segment and the horizontal

$$T_{ac}=23.56\text{kN} \quad T_{cd}=22.16\text{kN} \quad T_{de}=21.70\text{kN} \quad T_{eb}=24.48\text{kN}$$

Discussion

Since the horizontal component of the cable tension is constant, the tension will be a maximum where the absolute value of the slope is a maximum. For this cable it can be observed that this will occur between points e and b.

Unit 9　Yield Line Theory of Slabs

Introduction

Reinforced concrete design methods under the present ACI Code are based on the results of an clastic analysis of the structure as a whole, when subjected to the action of factored loads (ACI 9.2) such as $1.4D+1.7L$ where D and L refer to service dead and live loads. Actually the behavior of a statically indeterminate structure is such that after the ultimate moment capacities at one or more points have been reached, discontinuities develop in the clastic curve at those points and the results of an clastic analysis are no longer valid. If there is sufficient ductility, redistributions develop in the elastic curve at those points and the results of an elastic analysis are no will occur until a sufficient number of sections of discontinuity, commonly called "plastic hinges", form to change the struture into a mechanism, at which time the structure collapses of fails. The term "ultimate load analysis", as opposed to "clastic analysis", relates to the use of the bending moment diagram at the verge of collapse as the basis for design. Other than the provisions for redistribution of moments at the supports of continuous flexural members (ACI-8.4), the present ACI Code has as yet made no allowance for ultimate load analysis. The redistribution as described in ACI-8.4 has been presented and illustrated in Sec. 10.12.

The design of two-way slab systems has been treated in Chaps. 16 and 17. Even here the design moments are based on the elastic analysis of an equivalent frame, which is devised as a simple substitute for the elastic analysis of continuous plate systems.

The chief concern of this chapter is to develop the yield line theory for two-way slabs. Although not yet adopted by the ACI Code, slab analysis by yield line theory may be useful in providing the needed information for understanding the behavior of irregular or single-panel slabs with various boundary conditions.

General Concept

Although the study of flexural behavior of plates up to the ultimate load may date back to the 1920s, the fundamental concept of the yield line theory for the ultimate load design of slabs has been expanded considerably by K. W. Johanden. In this theory the strength of a slab is assumed to be governed by flexure alone; other effects such as shear and deflection are to be separately considered. The reinforcing steel is assumed to be fully yielded along the yield lines at collapse and the bending and twisting moments are assumed to be uniformly distributed along the yield lines.

Yield line theory for one-way slabs is not much different from the limit analysis of continuous beams. On a continuous beam the achievement of flexural strength at one location, say in the negative-moment region over a support, does not necessarily constitute reaching the ultimate load on the beam. If the section having reached its flexural strength can continue to provide a constant resistance while undergoing further rotation, then the flexural strength may be reached at additional locations. Complete failure theoretically can not occur until yielding has occurred at several locations (or along several lines in case of one-way slabs) so that a mechanism forms giving a condition of unstable equilibrium.

Yield line theory for two-way slabs requires a different treatment from limit analysis of continuous beams, because in this case the yield lines will nor in general be parallel to each other but instead form a yield line pattern. The entire slab area will be divided into several segments which can rotate along the yield lines as rigid bodies at the condition of collapse or unstable equilibrium.

Fundamental Assumptions

In applying the yield line theory to the ultimate load analysis of reinforced concrete slabs, the following fundamental assumptions are made:

(1) The reinforcing steel is fully yielded along the yield lines at failure.

(2) The slab deforms plastically at failure and is separated into segments by the yield lines.

(3) The bending and twisting moments are uniformly distributed along the yield line and they are the maximum values provided by the ultimate moment capacities in two orthogonal directions (for two-ways slabs).

(4) The elastic deformations are negligible compared with the plastic deformation; thus the slab parts rotate as plane segments in the collapse condition.

Methods of Analysis

There are two methods of yield line analysis of slabs; the virtual-work method and the equilibrium method. Based on the same fundamental assumptions, the two methods should give exactly the same results. In either method, a yield line pattern must be first assumed so that a collapse mechanism is produced. For a collapse mechanism, rigid body movements of the slab segments are possible by rotation along the yield lines while maintaining deflection compatibility at the yield lines between slab segments. There may be more than one possible yield line pattern, in which case solutions to all possible yield line patterns must be sought and the one giving the smallest ultimate load would actually happen and thus should be used in design.

After the yield line pattern has been assumed, the next step is to determine the position of the yield lines, such as defined by the unknown x in Fig. 1 (a) or (b). It is at this point that one may choose to use the virtual-work method or the equilibrium method. In the virtual-work method, an equation containing the unknown x is established by equating the total positive work done by the ultimate load during simultaneous rigid body rotations of the slab segments (while maintaining deflection compatibility) to the total negative work done by the bending and twisting moments on all the yield lines. Then that value of x which gives the smallest ultimate load is found by means of differential calculus. In the equilibrium method, the value of x is obtained by applying the usual equations of statical equilibrium to the slab segments, but the optimal position x is defined by the placement of predetermined nodal forces in typical situations, once derived, can be conveniently used to avoid the necessity of mathematical differentiations as required in the virtual-work method.

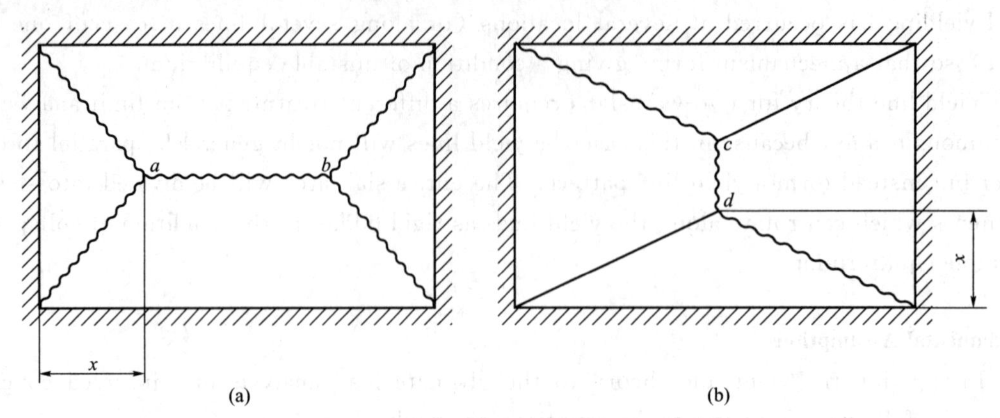

Fig. 1 Yield-line patterns of a simply supported rectangular slab

In the following sections, yield line analysis for one-way slabs is dealt with first in a manner similar to limit analysis of continuous beams. Then both the virtual-work method and the equilibrium method are presented and illustrated for two-way slabs.

Unit 10 Concrete Operations

Significance

Since a normal concrete mixture (i. e., concrete made with a normal Portland cement and subjected to normal handing operations and curing temperatures) generally takes 6 to 10 hr for setting and 1 to 2 days for achieving a desired strength level before formwork can be removed, the definition of early age includes on the one hand the freshly mixed concrete of plastic consistency and, on the other, 1 to 2-day-old hardened concrete that is strong enough to be left unattended (except continuation of moist curing).

The early-age period in the life of concrete is insignificantly small compared to the total life expectancy, but during this period it is subjected to many operations which are not only affected by the properties of the material but also influence them. For instance, a mixture with poor workability will be hard to mix; on the other hand, too much mixing may reduce the workability. It is beyond the scope of this book to describe in detail the operations and the equipment used, but engineers should be familiar with the sequence of main operations, their interaction with characteristics of concrete at early ages, and the terminology used in the field practice.

In general, the sequence of main operations is as follows: batching, mixing, and conveying the concrete mixture from the point where it is made to the job site; placing the plastic concrete at the point where it is needed; compacting and finishing while the mixture is still workable; finally, moist curing to achieve a desired degree of maturity before formwork removal. The operations described below are divided into separate categories only for the purpose of understanding their significance and the basic equipment involved; in practice, they may overlap. For example, in the truck mixing method, the mixing and transporting operations are carried out simultaneously.

Batching, Mixing, and Conveying

Most specifications require that batching of concrete ingredients be carried out by weight rather than by volume. This is because bulking of damp sand causes inaccuracies in measurement. Water and liquid admixtures can be batched accurately either by volume or weight. In many countries most concrete today is batched and mixed by ready-mixed concrete plants, where the batching is generally automatic or semiautomatic.

Abnormal handing and maturing characteristics of fresh concrete mixtures that are not uniform in appearance have often been attributed to inadequate mixing. Therefore, accurately proportioned concrete ingredients must be mixed thoroughly into a homogeneous mass. Depending on cost economy, type of construction, and amount of concrete required, the mixing operation can either be performed on-site or in a central off-site facility (ready-mixed concrete plant). On-site mixers of sizes up to 12 yd^3 which can be of the tilting or the non-tilting type, or the open-top revolving blade or paddle type.

Ready-mixed concrete is defined as concrete that is manufactured for delivery to a purchaser in a plastic and unhardened state. During the last 50 years of its worldwide development, the ready-mixed concrete industry has experienced tremendous growth. Most of the plants are equipped with automatic or semiautomatic batching systems and controls made possible by the use of microprocessors and computers. Truck mixing rather than central mixing is the commonly used method of mixing in the United States, although for the purpose of achieving better quality control the proportion of centrally mixed concrete increased from 15 percent to 27 percent, and truck-mixed concrete dropped from 83 percent to 71 percent during the period 1966—1977. Inclined-axis mixers of revolving drum type, either with rear or front discharge, are commonly used. In the past 10 to 15 years, for large or important work there has been a gradual change away from the prescriptive to performance or strength specifications. Also, increasingly, ready-mixed concrete producers are assuming technical responsibility for mix design and quality control.

Transportation of ready-mixed concrete to the job site should be done as quickly as possible to minimize stiffening to the extent that after placement, full consolidation and proper finishing became difficult. The causes and control of stiffening or loss of consistency, which is also referred to as slump loss, are discussed later. Under normal conditions there is usually a negligible loss of consistency during the first 30 min after the beginning of cement hydration. When concrete is kept in a slow state of agitation or is mixed periodically, it undergoes a small slump loss with time, but this usually does not present a serious problem for placing and consolidation of freshly made concrete within 1.5 hours. However, as discussed next, attention must be paid to possible delays in transporting and placing concrete under hot and dry weather conditions.

A summary of the most common methods and equipment for transporting concrete is shown as follows. In choosing the method and equipment for transporting and placing concrete, a primary objective is to assure that concrete will not segregate.

Placing, Compacting, and Finishing

After arrival at the job site the ready-mixed concrete should be placed as near as possible to its final position. To minimize segregation, concrete should not be moved over too long a distance as it is being placed in forms or slabs. In general, concrete is deposited in horizontal layers of uniform thickness, and each layer is thoroughly compacted before the next is placed. The rate of placement is kept rapid enough so that the layer immediately below is still plastic when a new layer is deposited. This prevents cold joints, flow lines, and planes of weakness that result when fresh concrete is placed in hardened concrete.

Consolidation or compaction is the process of molding concrete within the forms and around embedded parts in order to eliminate pockets of entrapped air. This operation can be carried out by hand ridding and tamping, but almost universally is carried out now by mechanical methods, such as power tampers and vibration that make it possible to place stiff mixture with a low water/cement ratio or a high coarse-aggregate content; mixtures of high consistency should be consolidated with care because concrete is likely to segregate when intensely worked. Vibrators should only be used to compact concrete and not to move it horizontally, as this would cause segregation.

Vibration, either internal or external, is the most widely used method for compacting concrete. The internal friction between the coarse aggregate particles is greatly reduced in vibration; consequently, the mixture begins to flow and this facilitates consolidation. One purpose of using internal vibrators (described below) is to force entrapped air out of the concrete by plunging the vibrator rapidly into the mixture and removing it slowly with an up-and-down motion. The rapid penetration forces the concrete upward and outward, thereby helping the air to escape. As the vibrator is removed, bubbles of air rise to the surface.

Internal or immersion-type vibrators, also called spud or poker vibrators, are commonly used for compacting concrete in beams, columns, walls, and slabs. Flexible-shaft vibrators usually consist of a cylindrical vibrating head, 3/4 to 7 in ches. in diameter, connected to a driving motor by a flexible shaft. Inside the head an unbalanced weight rotates at high speed, causing the head to revolve in circular orbit. Small vibrators have frequencies ranging from 10000 to 15000 vibrations per minute and low amplitude, between 0. 015 and 0. 03 in. (deviation from the point of rest); as the diameter increases, the frequency decreases and the amplitude increases.

External or form vibrators can be securely clamped to the outside of the forms. They are commonly used for compacting thin or heavily reinforced concrete members. While the concrete mixture is still mobile, vibration of members congested with reinforcement helps to remove air and water that may be entrapped underneath the reinforcing bars, thus improving the rebar-concrete bond. Precasting plants generally use vibrating tables equipped with suitable controls so that the frequency and amplitude can be varied according to size of the members and consistency of the concrete. Surface vibrators such as vibrating screeds are used to consolidate concrete in floors and slabs up to 6 in. thick.

Revibration of concrete an hour or two after initial consolidation, but before setting, is sometimes needed in order to weld successive castings together. This helps to remove any cracks, voids, or weak areas created by settlement or bleeding, particularly around reinforcing steel or other embedded materials.

Flatwork such as slabs and pavements require proper finishing to produce dense surfaces that will remain maintenance-free. Depending on the intended use, some surfaces require only strike-off and screening, whereas others may need finishing operations consisting of the sequence of steps described below, which must be carefully coordinated with the setting and hardening of the concrete mixture.

Screeding is the process of striking off excess concrete to bring the top surface to the desired grade. With a sawing motion a straight edge is moved across the surface with a surplus of concrete against the front face of the straight edge to fill in low areas. A darby or bull-float is used immediately after screeding to firmly embed large-aggregate particles and to remove any remaining high and low spots. Bull-floating must be completed before any excess bleed water accumulates in the surface because this is one of the principal causes of surface defects, such as dusting or scaling, in concrete slabs. When the bleed-water sheen has evaporated and the concrete is able to sustain foot pressure with only slight indentation, the surface is ready for floating and final finishing operations. Floating is an operation carried out with flat wood or metal blades for the purposes of firmly embedding the aggregate, compacting the surface, and removing any remaining imperfections. Floating tends to bring the cement paste to the surface; therefore, floating too early or far too long can weaken the surface. After floating, the surface may be steel-troweled if a very smooth and highly wear resistant surface is desired. When a skid-resistant surface is desired, this can be produced by blooming or scoring with a rake or a steel-wire broom before the concrete has fully hardened (but has become sufficiently hard to retain the scoring).

Concrete Curing and Formwork Removal

The two objects of curing are to prevent the loss of moisture and to control the temperature of concrete for a period sufficiently well above freezing, the curing of pavements and slabs can be accomplished by pounding or immersion; other structures can be cured by spraying or fogging, or moisture-retaining coverings saturated with water, such as burlap or cotton. These methods afford some cooling through evaporation, which is beneficial in hot-weather concreting. Another group of methods are based on prevention of moisture loss from concrete by sealing the surface through the application of waterproof curing paper, polyethylene sheets, or membrane-forming curing compounds. When the ambient temperature is low, concrete must be protected from freezing by the application of insulating blankets; the rate of strength gain can be accelerated by curing concrete with the help of live steam, heating coils, or electrically heated forms or pads.

Formwork removal is generally the last operation carried out during the early-age period of concrete. The operation has great economic implication because on the one hand, removing forms quickly keeps the construction costs low, while on the other, concrete structures are known to have failed when forms were removed before the concrete had attained sufficient strength. Forms should not be removed until concrete is strong enough to carry the stresses from both the dead load and the imposed construction loads. Also, concrete should be hard so that the surface is not injured in any way during form removal or other construction activities. Since the strength of a freshly hydrated cement paste depends on the ambient temperature and availability of moisture, it is better to rely on a direct measure of concrete strength rather than an arbitrarily selected time of form removal. For the safety of structures in cold weather, designers often specify a minimum compressive strength before concrete is exposed to freezing. In hot weather, moisture from fresh concrete may be lost by evaporation, causing interruption of the hydration and strength gain process; also surface cracking due to plastic shrinkage may occur.

Words and Phrases

1	overlap	*n*.	重叠
2	simultaneously	*a*.	同时的
3	batch	*n*.	配料
4	bulk	*n*.	湿胀性

5	paddle	*n*.	浆状
6	Inclined-axis	*n*.	斜轴
7	consolidation	*n*.	捣实
8	segregate	*v*.	离析
9	deposit	*v*.	沉淀
10	compaction	*n*.	压实
11	tamp	*v*.	夯实
12	penetration	*n*.	贯入
13	slab	*n*.	平板
14	evaporate	*v*.	蒸发
15	float	*v*.	持平
16	sheen	*n*.	光泽

Exercises

1. Translate the following into Chinese

(1) In general, the sequence of main operations is as follows: batching, mixing, and conveying the concrete mixture from the point where it is made to the job site; placing the plastic concrete at the point where it is needed; compacting and finishing while the mixture is still workable; finally, moist curing to achieve a desired degree of maturity before formwork removal.

(2) When concrete is kept in a slow state of agitation or is mixed periodically, it undergoes a small slump loss with time, but this usually does not present a serious problem for placing and consolidation of freshly made concrete within 1. 5 hours.

2. Translate the following into English

(1) 振捣器的快速贯入使混凝土向上或向外运动，因此帮助空气排除。

(2) 混凝土的再振捣帮助消除任何裂缝、空隙或沉降或析出产生的薄弱区，尤其在配筋或其他埋入材料周围。

Unit 11 Future Trends in Construction

Construction is the translation of a design to reality. This is an engineering problem that is just as exacting and challenging as the design, and must receive the same careful study as the design in order that the structure will perform as it was intended to and the work is completed within the required time at a minimum cost.

Apart from changes for strictly esthetic reasons, design will become influenced more intensely by such considerations as economics, ecology, energy, and other facets of life, some of which are on the way to reaching crisis status. This will tend to lead to many innovations and tax the designers' ingenuity. With the gradual realization of the limitations of natural resources and the global need for conservation it will become necessary to maximize efficiency in all areas.

IT is interesting to note that the close approach and interrelationship between construction and manufacturing is gradually eroding what used to be previously a well-demarcated area. High costs, problems with organized labor, and construction in northern climates under severe weather and climate conditions have given special impetus to prefabricating as many building components as feasible. Thus actual field erection is reduced appreciably, permitting faster completion and increasing efficiency and profitability of the project.

The future has many surprises in store for us. We are already thinking of possibilities which are still only a gleam in the experts' eyes.

Excavation will be performed for large projects by applying nuclear methods with special radiation-proof equipment and methods.

Prefabrication will be much more advanced and a universal standardized code based on using metric modular units will be applied on a global basis. Not only will prefabrication permit assembly of building components manufactured in different locations at distant sites, but it will also be engineered so that various appliances and equipment equally standardized, can be used anywhere, built-in if so desired, and with no worry about not being able to find spare parts or repair facilities locally in case of mechanical failure.

Many daily functions will become automated and computer-controlled, especially in residential construction. Lighting, heating, kitchen services, and many other functions will be timed to suit personal requirements and the complete integrated program win be capable of being triggered or adjusted by remote control commands given by telephone. General news, private messages, pictures, and even games will be projected on special wall and ceiling panels by means of built-in projection equipment. Many other functions in business and industry will be treated similarly. Miniaturization of formerly bulky items, thanks to modern electronic know-how, including microcircuitry and solid-state technology, will not only reduce size and space requirements of many building components control elements, but in view of the reduced power requirements will also help to conserve energy. All this will require special, facilities and installations and will add greatly to the complexities of both design and construction.

Our changing energy needs and sources will affect building design greatly. Solar power will at first tend to be developed on an individual basis, with various houses or buildings having their own installations. This may be followed by complete solar pr stations combined with underground hot water storage reservoirs and other ancillary facilities. Wind power win be utilized more in wind power stations. These, however, will only find efficient use in regions having sufficient wind currents, such as coastal areas and some mountainous regions. Coal, a longtime pariah among energy sources, will find new prominence, since the scrubbing and purification required make it ecologically acceptable will now be economically justified in comparison with the escalating costs of petrochemical and nuclear energy. Nuclear power stations may become commonplace for general power generation, but a lot of equipment will be powered by mini-nuclear power units, all mounted in accident-proof sealed units, which will outlast the equipment and be capable of being reused again and again.

Construction methods will change in all areas and manual labor will tend to disappear with more and more emphasis on mechanized operations. Thus there will be a demand for more specialists, primarily those having technical backgrounds.

Research and experimentation by inventive minds, both in the field and in the laboratory will lead us into an era of progress in technology which will open new vistas and enable us to use our vast technical knowledge for the good human beings throughout the world.

Unit 12 Scheduling and Control of Construction

On very large projects to be constructed for government agencies or large private institutions, fiscal policies often impose very rigid limitations on how much of the total commitment will be made available in each fiscal year, and usually require that the money be spent during the year that it is budgeted. Failure to spend the money in time could mean loss of the money.

Some projects are of such a size that few, if any, contractors have the financial and personnel capacity to undertake them. Those contractors physically capable of doing the job may feel it is too much of a risk to take on one project alone. They would rather divide that amount of risk among three or four different projects. Too large a project may also drain the labor market and cause "labor piracy" inefficiency, and premium cost. In order to obtain enthusiastic bids in sufficient numbers, large construction jobs are often divided into several smaller sections and separate bids solicited for each phase.

In solving the above problems, construction management problems are created that must be skillfully solved. Those operations that are prerequisite to others must be completed before the next operations are started. Parallel and supporting operations must be phased in so as to be completed without holding up the critical work or adding time to the intermediate completion dates. Certain items that require long fabrication and/or delivery dates may have to be ordered very early and contract documents must allow and ensure such early purchases.

The most popular tool for controlling and dovetailing these operations is the critical-path chart. By showing all planning, procurement, fabrication, construction, and testing operations as vectors to a time dimension, these operations are so arranged that all operations follow immediately after prerequisite operations and as concurrently as possible. When the chart is properly done, one call tell the minimum construction time required, critical schedules that must be held, slack time in operations not on the critical path, effect of speeding certain operations, etc. This chart is no better than the input it receives, the contract documents governing the work, or the construction managers implementing it.

Design is prerequisite to all operations. Usually the time required for completion of the entire design will set back the construction schedule excessively. It is therefore necessary to carry out the design concurrent with construction with only a limited lead-time.

Once a job is divided among several contractors, each having himself interests to promote, it becomes necessary to schedule and to cover the work by document and field management in such a way that interferences between operations are eliminated or minimized. There are, also, the interests of the client, such as an airport authority, which must maintain operation during construction; transportation and municipal agencies responsible for the streets and highways that the contractors must use; and, of course, the neighbors.

If the interaction of these problems is left to be solved "in the field", the result will be litigations, delays, and excessive extra costs. The architect or engineer must plan the construction procedures in considerable detail, with the help of construction consultants, if necessary.

It is almost always necessary, especially in vertical construction, that the contractors use the same site area of the job. Hopefully this can be managed in a chronological order; often it is concurrent or at least with a time overlap. This division by time and area must be clearly defined, taking into consideration the realistic needs of the contractors involved. Very often equipment such as cranes, hoists, personnel elevators, utility lines, and permanent installations temporarily used for. Construction are required by more than one contractor in sequence, or concurrently.

If each contractor can have his own equipment fully under his own control, this is the ideal condition. Often space requirements prohibit this. There is obviously an unnecessary cost that always accrues to the owner if each contractor erects and dismantles the same tower crane on the same spot as he performs his particular phase of the work. Economy and site efficiency can best be served by specifying equipment capable of performing all phases and operations of the work, scheduling the hours of use to the contractors involved, and arranging for an equitable sharing of cost through a rental or allowance arrangement.

Inspection is more difficult and more critical than ever. The industry is using new materials, new techniques, higher allowable stresses, and lower safety factor (ultimate strength design). There is more room for error and less allowance for it. All too often, the only qualification for an inspector is that he is willing to work for a very low salary. This is a very dangerous procedure for modern sophisticated construction. Inspectors must have some training in construction technology, must be given special instruction by the architect and engineer in charge, and must have open lines of communication to their superiors and to the contractor.

The design architect and engineer must visit the job often to check on and to instruct the resident staff. In some instances, when specialized knowledge is not available in-house, it is necessary to obtain the field assistance of construction consultants, manufacturers' technicians or engineering specialists.

The serious inspector will spend any free time he gets on the job studying the plans and specifications and reading relevant technical books and papers. A little homework would not hurt either. But remember, if this type of intelligent conscientious person is required on the job, the remuneration must be commensurate. Very often inspection degenerates into enforcing the letter of the law without regard to the way in which the instructions of the contract documents are affecting the work. The inspector is, and must be, a policeman. To be really effective, however, he must know the limits within which he may make his own decisions and must know when to request help from the design engineer. In his turn, the design engineer must clearly instruct his inspector in the field as to what is inflexible, what is flexible, and to what extent, and when to call for help.

Unit 13 Contractors' Management Game

The majority of building and civil engineering contracts in the UK construction industry are offered through a process of competitive tendering. This is the process whereby the client or his representative prepares the contract documents including drawings, specifications and bills of quantities, and arranges that two or more contractors submit a tender or price . for the contract. Based on this tender and other relevant information the client is free to choose which contractor should be awarded the contract.

A contractor who is tendering has to prepare an estimate for the cost of the work based on the client prepared contract documents and any other investigations he wishes to make. To this cost estimate the contractor adds a mark-up which is an allowance for profit and company overheads not directly associated with the particular job. The sum of estimate and mark-up is the tender, or bid.

Tender=estimate of contract costs+mark-up

(includes profit margin and allowance for company overheads)

The actual cost of executing the contract incurred by the successful contractor will be at variance with his original estimate. The contribution, i. e. the margin remaining for company overheads and profit, is the tender value less the actual cost.

Contribution=tender−actual cost

Company profit is the sum of the contributions from all contracts won less the company's overheads.

Company profit=contributions from all contracts won−company's overheads

Usually such a figure as this would he related to some time period, for example, one year.

Almost all contracts in the construction industry allow a series of interim payments to be made to the contractor. The amount of these interim payments is based on a valuation of the work already completed less a retention held by the client until the completion of the contract. Typical retention in the UK construction industry is 5% to 10 % of contract value. Typical achieved profit margins are less than 10% and are often less than 5 % of contract value. Thus it is not unusual for a company to find all its profit locked up in retentions.

In the game each of the participants plays the role of a company. Each company obtains all its work through the competitive process. In the real world situation, while many companies are dependent on the competitive market for most of their work, few rely on it completely and make up the remainder of their turnover with negotiated contracts or other diversifications.

The game does not attempt to simulate the estimating procedures of a contractor, but enters the process with control decisions on such questions as "Do we want this job and, if so, at what figure are we prepared to tender?" These important decisions are made against a background of uncertainty, the uncertainty arising from the difficulties in predicting the behaviour of competitors and the knowledge that the cost estimate is inaccurate.

The cash flows of a company in the game are calculated on the given predetermined retention conditions, payment delays from clients and credit limits from suppliers. Although these important variables in controlling cash flow are fixed, the company in the game has control of its cash flow through the number of contracts it takes on, the turnover mix (i. e. the turnover made up of all small short jobs or large jobs of long duration) and the profit levels at which the company operates.

A difficulty in operating the game has been to decide what should be done with companies in serious cash difficulties. In real life, of course, these companies would be declared bankrupt. Would be wound up and their place in the industry would be filled by expansion of other companies and the creation of new companies. In the game situation at least three possibilities exist.

(1) declare the company bankrupt and have them leave the game.

(2) declare the company bankrupt and allow the team to set up a new company.

(3) do not have bankruptcy and allow the company to continue struggling.

Unit 14 The Construction Process May Be Automated In the Future

Automation has changed many aspects of life. At one time, only wealthy individuals could afford an automobile; now, thanks to automated processes, cars are assembled by efficient robots and churned out by the thousands each day in sprawling factories. A University of Southern California (USC) professor hopes his National Science Foundation-funded research will revolutionize architecture and the construction sector in much the same way. His work with an automated process called contour crafting suggests that, in the future, erecting a house or office tower may involve very few human beings. Contour crafting uses a machine to execute a computer aided design (CAD). The machine-called a contour crafter operates on a gantry system fitted with rails and employs a nozzle that deposits dozens of layers of concrete in a specified layout to build walls of almost any shape or size, says Behrokh Khoshnevis, Ph. D. , a professor in USC's industrial and systems engineering department. It even digs and places a concrete slab foundation.

To achieve a structurally sound wall, the contour crafter lays down two strips of concrete along the path of the wall and fills the space between the strips with an insulating concrete mix. It also uses trowels and other standard tools to give surfaces a smooth finish that Khoshnevis says is superior to hand finishes. The apparatus can also grasp and install such elements as roof beams which are subsequently topped with concrete-and steel reinforcing members for load bearing walls and can leave space for such necessities as electrical wires, water pipes, and ventilation shafts. The contour crafter can also leave openings in the walls for windows and doors, which would later be installed manually.

The machine operates under a canopy to ensure that it is not interrupted byinclement weather, and it can complete a modest single family residence in as little as 24 hours. According to Khoshnevis, a contour crafter can also create more complex structures, such as skyscrapers and multistory apartment and office complexes. He says that the machine can be programmed to "climb" the structure as it builds.

It can also meet whatever structural requirements are necessary for tall buildings, he says. "You can put a lot of reinforcement in the walls or make them thicker," says Khoshnevis. "The contour crafter does what you specify in the design. " And providing additional reinforcement won't slow down the process, he says, since the contour crafter can concurrently reinforce a wall in one area and deposit concrete in another part of the structure. Multiple contour crafters on parallel gantry rails could create a wider structure. Mobile robots with contour crafter nozzles also could build more complicated structures. In this take on the concept, rails would not be used. The robots would have their own concrete . The contour crafter uses automated processes to erect structures rapidly. Behrokh Khoshnevis, Ph. D. , a professor at the University of Southern California, believes this technology has the potential to revolutionize both architecture and construction. Research is under way to develop a quick drying concrete mix for the machine.

Mixing tanks and would be programmed to refill the tanks when they were running low. Once the contour crafter is in operation, the system can be monitored remotely via a webcam. "The machine will have a lot of self-diagnosis capabilities and could report problems it encounters, such as inconsistencies in material, extreme wind conditions, and so on," says Khoshnevis.

According to Khoshnevis, a contour crafter has built a full-scale concrete wall section in the laboratory; but thus far it has not created a large-scale structure, such as a house. He hopes that milestone will be achieved by the end of the year. In the meantime, improvements are being made to the system. For example, several companies are in the process of developing concrete mixes that dry very quickly and that can be used by the machine. "In the near future, concrete will be able to set in about twenty-four hours," Khoshnevis says.

Contour crafting could have a significant effect on refugee populations and developing nations. Khoshnevis believes that automating the building process could bring decent, affordable housing to millions of people worldwide. "Today, most people in third world countries all have shoes and clothing, but twenty or thirty years ago, they didn't have those either," he says. "These days, they have the kind of shoes that we [in America] wear because automation made it possible." According to Khoshnevis, it's just a matter of time, before automated construction eradicates slums and ghettos. By the same token, refugees who currently are forced to live in substandard conditions-many intents may someday be housed temporarily in structures built by contour crafting.

According to Khoshnevis, the Construction Engineering Research Laboratory-a division of the U. S. Army Corps of Engineers' Engineer Research and Development Center is interested in the contour crafting technology. "They're talking about using it to build not just structures such as offices but barriers and barracks as well," he says.

Contour crafting would help the military construct sophisticated buildings much more quickly. In a further application, the National Aeronautics and Space Administration (NASA) is considering contour crafting as a means of erecting structures on the moon. In collaboration with NASA's Jet Propulsion Laboratory, Khoshnevis's lab is designing a contour crafter made from three mobile components that could be flown into space. Once on the moon, two of the individual components would assemble themselves to form the contour crafter, he says. Instead of using concrete as the structural material, the third component would excavate and collect lunar regolith that, after being melted in microwave ovens powered by solar cells, would be fed to the contour crafter. The melted material would then be extruded, and because the moon's atmosphere lacks oxygen, the deposition of the resulting material would be "very consistent and solid ... you're not going to deal with the porosity and oxidation issues commonly seen in volcanic lava on Earth," Khoshnevis says. The machine could use the material to construct various types of structures-such as residences, laboratories, and observation platforms-the designs for which would be sent to the crafter by e-mail. "In a period of a year, it could build a small town," says Khoshnevis. If reinforced structures were needed, glass fibers could be created from in situ lunar materials, he says. NASA's Marshall Space

Flight Center is using the use contour crafter in building trials with simulated lunar materials.

Contour crafting is still prohibitively expensive between $500000 and $1 million for a machine that would create the average [2000 sq. ft. (185.8 sq. m)] house, says Khoshnevis. However, he expects that once the technology catches on the cost of the machine will be significantly reduced. Ultimately, erecting a structure by contour crafting may cost as little as one-fifth the amount incurred in conventional construction. "This is very much in line with such other manufactured goods as shoes, clothing, and cars, which would typically cost four to five times as much if created by hand," Khoshnevis says. He foresees a world in which architects and prospective homeowners could create and customize CAD designs for residences and then build them with the contour crafter. "The possibilities are endless," he says.

Unit 15 FIDIC Tendering Procedure

The International Federation of Consulting Engineers (FIDIC) published the first edition of Tendering Procedure in 1982. The first edition primarily addressed procedures which FIDIC recommended for the selection of tenders and the preparation and evaluation of tenders for civil engineering contracts. The document reflected the provisions of the then current (third edition) of the FIDIC Conditions of Contract (International) for Works of Civil Engineering Construction.

This document presents a systematic approach for tendering and awarding of contracts for international construction projects. It is intended to assist the employer/engineer to receive sound competitive tenders in accordance with the tender documents so that they can be quickly and efficiently assessed. At the same time, an effort has been made to provide the opportunity and incentive to contractors to respond easily to invitations to tender for projects which they are qualified to implement. It is hoped that the adoption of this procedure will minimize tendering costs and ensure that all tenders receive a fair and equal opportunity to submit their offers on a reasonable and comparable basis.

Prequalification of Tenders

Prequalification documents should give information about the project, the tendering procedure and the prequalification procedure. They should also specify what data is required from contractors wishing to prequalify. The documents are prepared by the employer/engineer and will normally include the following: Letter of invitation to prequalify; information about the prequalification procedure; project information; prequalification application.

The employer/engineer should publish a notice inviting interested contractors to apply for prequalification documents, stating that tender documents will be issued only to a limited number of companies/joint ventures selected by the employer/engineer as having the necessary qualifications to perform the work satisfactorily.

The notice should be published in appropriate newspapers and journals to give sufficient publicity according to the particular circumstances of the project. The notice may also be issued to financing institution representatives, if relevant, and to government agencies responsible for foreign trade so that the international community receives timely notification of the proposed project and instructions on how to apply.

The notice should be reasonably brief and where feasible contain: (1) Name of the employer; (2) Name of the engineer; (3) Location of the project; (4) Description of the project and scope of work; (5) Source of finance; (6) Anticipated program (I. e. award of contract, completion and any other key dates); (7) Planned dates for issue of tender documents and submission of tenders; (8) Instructions for applying for prequalification documents; (9) Date by which applications to prequalify must be submitted; (10) Minimum qualification requirements and any particular aspects which could be of concern to prospective tenders.

The period between the notice of invitation to prequalify and the latest date for the return of completed applications should not be less than four weeks.

The employer/engineer should evaluate the prequalification applications to identify those companies/Joint ventures whom they consider to be suitably qualified and experienced to undertake the project, if the resulting list, after those firms who were found unsuitable have been excluded, exceeds six potential tenders and there are no special regulations or conditions imposed on the employer, the selection procedure should be continued to eliminate the less well-qualified in order to arrive at no more than six.

When the list of selected tenders has been prepared, successful applicants should be notified and requested to confirm their intention to submit a tender. This should ensure, as far as possible, an adequate number of competitive tenders. If a potential tender wishes to drop out at this stage, the next best-placed should be invited and asked to confirm as above. Following this, all applicants should be notified of the list of selected tenders without giving explanation for the decisions.

Obtaining Tenders

The tender documents prepared by the employer/engineer will normally include the following:

—Letter of invitation to tender.

—Instructions to tenders.

—Tender form and appendices.

—Conditions of contract (Parts Ⅰ and Ⅱ) together with any requisite forms.

—Specification.

—Drawings.

—Bill of quantities or schedule of prices.

—Information date.

—List of additional information required from tenders.

Instructions to tenders should be prepared by the employer/engineer to meet the particular requirements of individual contracts. The purpose of the document is to convey information and instructions that will govern the preparation, submission and evaluation of tenders.

When determining the tender period, the employer/engineer must ensure that adequate time is available for tenders to prepare their tenders taking into account the size, complexity and location of the project in question.

Tenders should be notified of the number of copies of their tender that are required, stipulating that one set of the documents should be, clearly marked "Original Tender" and the others (which should be photocopies) marked "Copy" and that, in the events of discrepancy, the "Original Tender", shall take precedence.

The instructions to tenders should state that the employer does not bind himself to award a contract to any of the tenders.

Tenders should be advised of the source of finance and related conditions. Where tenders are required to provide financing they should be instructed to provide information as to source of finance and the conditions which will apply.

Specific instructions should be given concerning the currencies to be used in the preparation of the tender. Tenders should also be advised in which currency/currencies payments will be effected.

The requirements for a tender security, if any, will be determined by the circumstances of each project. If a tender security is required, a form should be included in the tender documents. The amount and currency/currencies of the security should be stated. In all cases the surety or sureties must be satisfactory to the employer, if a tender security has been requested, any tender which has not been so secured will specification.

Specification

The specification will define the scope and technical requirements of the contract, including any requirements for training and the transfer of technology. The quality of materials and the standards of workmanship to be provided by the contractor must be clearly described, together with requirements for quality assurance to be performed by the contractor and the required safety, health and environmental measures to be observed during the executions of the works, the extent, if any, to which the contractor will be responsible for the design of the permanent works should also be specified. Details should be included of samples to be provided and tests to be carried out by the contractor during the course of the contract. Any limitations on the contractor's freedom of choice in the order, timing or methods of executing the work or sections of the works would be clearly set out and any restrictions in his use of the site of the work, such as interface requirements with other parts of the work, or provision of access or space for other contractors, should be given.

The specification shall promote the broadest possible competition and as for as possible follow international standards such as those issued by ISO.

Drawings

The drawings included in the tender documents should provide tenders with sufficient detail to enable them, in conjunction with the specification and the bill of quantities, to make an accurate assessment of the nature and scope of the works included in the contract. The drawings should be listed in the specification.

Evaluation of Tenders

Following the opening, tenders should be checked by the employer/engineer to establish that they are arithmetically correct, are responsive without errors and omissions and consistent with the invitation to tender.

The evaluation of tenders can generally be considered to have three components. The components may include: Technical evaluation; Financial evaluation; General contractual and administrative evaluation.

Award of Contract

The employer will normally seek to award the contract to the tender submitting the lowest evaluated responsive tender. The award must be made during the period of tender validity or any extension thereto accepted by the tenders. The contractor should normally be required to sign a contract agreement with the employer. The employer/engineer should prepare the contract agreement which should include the following documents:

—Letter of Acceptance and Memorandum of Understanding.

—Letter of Intent (if applicable).

—The tender.

—Conditions of contract.

—Specification.

—Drawings.

—Bill of quantities.

—And such other documents that are intended to form the contract.

Unit 16 Introduction of Industrialized Construction

Introduction

The industrialization of building affects each person in the total building process; in the struggle for survival in our industry—as it moves towards a more "industrial" method of operation, each one of us constructs an image of that future that reflects his own present-day view point; he defines the main characteristics of that future in or terms of his own priorities.

The papers that form this group aptly illustrate this divergence. Each author, responding to the challenge to view the immediate past, the present and the future of industrialization, brings out points that are pertinent to his own experience; in so doing, he also reveals his own preconceptions, and his own scale of values.

The purpose of these notes is to suggest some generalities about the industrialization of building, that will enable the other papers to be viewed as parts of a whole, rather than as apparently conflicting statements about an ill defined subject.

Definitions

We must start by being clear about some of the terms we use, here are some definitions and explanations.

Industrialization. a productive method, based on mechanized and/or organized processes of a repetitive character requiring continuity... Note in this definition the two facets: (a) methods and processes; (b) repetition and continuity. I do not endorse the use of the term "industrialized" as applied to building-and as used in the title of this group of papers—since it would seem to imply that there is some other kind of building which is not industrialized. Non industrialized building does not exist; even the most traditional forms of construction use some industrially made products (bricks, nails, glass and the like). Indeed, these products are mass-produced. What we are concerned with, therefore, is an evolutionary process; when we refer to industrialization in building, we mean "increasing the amount of industrialization..."

Now, the industrially made products of traditional building are small and simple. They can be used in almost any kind of building-in other words, they can be mass-marketed into the traditional building industry, with all the variety that it contains. To assemble them, however, many manual (i. e. non-industrial) operations are required. Therefore, when we call for more industrialization, it is likely that we are also calling for the industrial manufacture of more complex products-designed to be assembled industrially. As we will see, This soon becomes a major departure from traditional building.

System. As soon as industrialization is discussed, the concept of system raised. A system is a set of objects with relationships between them and between their attributes... Note again two significant facets: (a) set of objects; (b) relationships.

Environment. Any discussion of systems immediately brings up the concept of environment. The environment of a system is: those objects outside the system (a) which are influenced by a change in the system or (b) a change in which influences the system.

In building terms, the building industry is a system. Its members are the "objects", the network of contracts and communications between them are the "relationships". The environment of the building industry system includes political, social and economic interests of many sorts.

The Systems Approach

If we now associate these two concepts-evolutionary industrialization taking place within the building industry system, we can begin to understand why the problems are complex and why it is so difficult for one or two members of the industry to make a significant impact on the degree of "industrialization" in building, when they "go it alone".

Traditional building products, I stated, are often mass-produced, but nearly always being simple-require manual assembly processes. Industrialization, I have also suggested, implies a change in this situation, leaning to the use of more complex products where the "complexity" is specifically aimed at simplifying the assembly processes. This, in itself, involves some major industrial design type activities; in addition, there is a production consideration that must not be overlooked: introducing new, easy to assemble products would be pointless if these more complex products—though easy to assemble—had to be hand made!

Now there is a relationship between the size (or complexity) of a product, the variety of situations in which it can be used and the number models that must be on offer. For a given variety: bigger products need more models; for a given number of models, bigger products mean less variety, etc. From this, we can immediately deduce that the call for industrialization and its more complex products, interferes with the present delicate balance between the kind of industrialized product being made for building and the kind of variety that the building industry traditionally demands.

In other words, a technological change—the limits of which can be controlled by one or two members of the building industry, immediately has far reaching consequences. To produce complex products industrially, i. e. to offer a reasonably restricted number of models, implied that the number of situations in which they are used is appropriately restricted; this, in turn, implies that the design of the building is somehow adjusted to these products, and also the decisions to purchase the buildings are affected. We see that the technological change has repercussions that reach right through the total building process. Once we remember that the building industry is a system, this becomes less surprising; the consequences are difficult to deal with, even if they are less surprising.

In effect, this is an example, one example, but an important one of how any partial change of building practice immediately produces a kind of chain reaction. A good model of the building industry would be a playground climbing net; change the position of one climber and all the others have to readjust their positions as the stresses redistribute themselves; indeed, the support ropes all change directions too, interacting differently with the supports themselves.

When the systems approach is discussed, one is recognizing that the building industry is a system and also recognizing that any one change should be accompanied by a systematic prediction of, and positive action on, the consequences elsewhere in the system. When systems building is discussed, one is referring to situations in which a limited systems approach is adopted, in which "building systems" (named ventures often of a proprietary nature) are proposed; these "building systems" are still subsystems of the building industry as a system; one hopes that the subsystems in question control enough factors to be able to make a success of the individual changes—industrial changes—within it. Building systems often run into trouble, as we all too well know—usually it is because of some unhealthy relationship between them and their environment (remember that the environment of a building system is the building industry and its environment). The problem area is usually marketing.

In the area of marketing, many building system sponsors behave with extraordinary naivete conducting only the most generalized market studies. It is essential to start from the right premise and distinguish between the "need" and "effective demand"; "need" is a social measure of shortage (e. g. 26 million dwellings in the U. S. A. between 1968 and 1978, 365000 dwellings per annum in the U. K. in the'60s); "effective demand" is a statement of the numbers of buildings for which contracts can be placed with the building industry. Effective demand is often very different from need; money can be made or lost against effective demand; as for need—I suppose only politicians or sociologists careers can be built on it.

In short, the systems approach to industrialization requires that market be studied thoroughly—as part of the systems analysis of the environment of the proposed system.

Performance

Any discussion of industrialization brings up the concept of performance; supposedly the description of "effective demand" in performance terms (i. e. without any reference to the nature of the physical solution) opens the door for innovation. This is undoubtedly true up to a point since performance terminology avoids certain descriptive constraints couched in terms of such and such a material, however we must be extremely cautious for several reasons; notably (a) it is extremely difficult to write good performance specifications. Indeed, it is at present impossible to write performance specifications without some allusion to the nature of the physical solution, and (b) it is not yet possible to deduce the physical solution from a performance specification (we can postulate a solution and verify its performance characteristics against the performance criteria as specified).

Skills

To proceed in performance terms implies a new set of design skills and design procedures; no longer is it possible (for technological and administrative reasons) for the design process to proceed in the traditional one-way multi-step way, starting with the design brief and finishing with site construction. Instead, it will proceed on several fronts, simultaneously, with systematic user-requirements studies leading to performance statements, with product designs accompanied by systematic assembly studies, with industrial engineering proposals leading to the design of new methods and new equipment. In other words, these will be new roles requiring new skills and new relationships; these skills will—at one and the same time—require a mixture of a systems engineering ability to view a whole problem area, and a scientist's skills in isolating part of that problem and finding new solutions with predictable behavior.

Industry Structure

To exercise these roles will require, as I have mentioned, new relationships between the participants; however, I do not believe that the giant building corporation is the best framework within which this work can take place. The traditional building industry has many merits that should not be overlooked: for example, its versatility: the entrepre-neurial character of its members, etc. Some way must be found to minimize the short comings of the industry; however, I suggest that its present dispersion and discontinuity should be replaced by coordinated management functions such as systematic communications, restructured responsibilities and so on. Note that coordinated management is not dependant on commercial integration of the various bodies. It is perfectly possible to have coordination between economically and commercially independent bodies; it is more difficult to have coordination when a state of potential conflict of interests exists.

Conclusion

To summarize, we have defined industrialization in terms of processes and systems in terms of relationships. To increase the amount of industrialization, we may have to change the nature of the products with which we build, implying a new attitude to the market for our services. The performance concept is one aspect of a change in defining the relationship between user requirements and the physical solution, but it immediately has repercussions on the roles of the various participants in the design process. However, the ensuing structural changes in the industry need not lead to the creation of giant building corporations; instead, management tools, (communications in particular) can prepare the traditional building industry for the systems approach that must accompany industrialization.

Unit 17 The Commercial Property Investment of Green Building

In the current debate on global climate change, green building investment is increasingly considered by experts and institutional investors to act as vehicle for environmental impact mitigation and for achieving energy efficiency, carbon reduction, and corporate social responsibility. The pressure to shift to green building is anchored by the rising evidence that the building sector is a major consumer of resources and energy, taking about 44% of the society's total material use and a large proportion of more than 50% of primary resources. For example, energy consumption by buildings in Canada, UK, and the US is placed between 30% and 50% of the country's total energy demand. Commercial properties contribute significantly to this problem. Commercial buildings (offices, retail, and industrial) consume close to 20% of the total energy consumption. In Malaysia, commercial buildings alone account for about 32% of the total energy consumption. There are also increasing body of studies indicating that green buildings could contribute to 30% to 50% reduction of total energy use, 35% reduction of carbon dioxide (CO_2) emission, 40% reduction of water usage, and 70% savings on waste output.

Malaysia has joined in the green building chase as increasing evidence emerge that green buildings are environmentally sustainable and can enhance productivity, lower market risk, and save cost over their operational life. To realize these benefits, Malaysia has developed some pro-green building policy measures. Among such measures are Malaysia green building rating system known as Green Building Index (GBI), National Green Technology Policy (NGTP), Low Carbon Cities Framework and Assessment System (LCCF), Malaysian Carbon Reduction and Sustainability Tool (My CREST), and Minimum Energy Performance Standards. Moreover, to demonstrate leadership and commitment to green building, Malaysian government has retrofitted four of her iconic public buildings [the Diamond Building in Putrajaya, the Kuala Lumpur Securities Commission Building, Low Energy Office (LEO), Green Energy Office (GEO) Building, Green Tech Malaysia, Green Technology and Water Building] into green buildings. Malaysia has also provided some corporate green tax incentives for companies, but the incentives are insufficient to attract investors.

Despite these policies and measures, the market for green building in Malaysia and Southeast Asia is still overcast and uncertain; as a result, potential investors are still holding back. Consequently, Malaysian building developers have been cautioned to take a respite and think deeply before investing in green buildings. This sentiment was also echoed by Eichholtz et al., who noted that real estate developers and institutional investors are justifiably not sure on how far to go in the green building investments. This is mainly due to the fact that existing economic justifications for the development of sustainable buildings rely mostly on anecdotal evidence.

It may be under this cloud of uncertainty that causes, among nations, Malaysia to not yet be in the forefront of green building leadership. In an international comparative study on "Green Investment Gap", to determine the countries that invest heavily in energy innovation and green investment, Japan and Finland were ranked the highest, sequentially followed by Korea, France, Demark, Norway, Sweden, US, Italy, Germany, UK, Spain, and Ireland. While the neighbouring country Singapore was on the list, Malaysia did not make the ranking. Consequently, green building supply and investment are still low in Malaysia despite the availability of the market.

Green building cost, availability of green building incentives, and green building skills are fundamental barriers to green building development particularly in the developing countries, including Malaysia. Research relating green building development and investment to these areas is very limited. Leading countries and cities in green building are those who are reducing green building cost through the significant provision of green building incentives and have developed high green building skills in designs, construction, maintenance, and technologies in energy efficiency, water efficiency, and material use efficiency. As the International Labour Organization pointed out, there are shortages of skills in the green building sector due to the changing requirements. Skills that were previously satisfactory are no longer meeting the present requirements in green building. For instance, a study by Aliagha et al. discovered a huge green building skill gap between the current and future skill requirement for energy-buildings in Malaysia. Specifically, Aliagha et al. found a wide gap between the current and future green building skill requirements for (1) efficient light system design with controls; (2) efficient passive wall, roof, and floor design; (3) efficient passive wall, roof, and floor insulation installation; (4) efficient passive window glazing design and installation; (5) efficient solar photovoltaic panel design and installation; (6) energy-efficient HVAC system design; (7) energy-efficiency diagnosing and auditing; (8) carbon capture and storage; and (9) energy-efficient maintenance, especially HVAC system maintenance. In many respects, the insufficient availability of green building skills represent a major obstacle to what can be achieved in green building. In the absence of sufficient professional skills in green building, the performance of a building planned to be green may be severely compromised.

There appears to be no empirical evidence evaluating the motivating factors of green commercial property investment in relation to green building incentives and green building skills, especially in the developing nations. Current studies on green building seem to focus on green residential buildings government and institutional green buildings, and energy efficiency. Even though green commercial properties are gradually becoming areas of research interest, available studies focusing on green commercial properties seem to focus mainly on green building certification, energy efficiency, eco-labelling, green building, and productivity, without specific attention to the interdependent factors that underlie the supply for green commercial property investment. Moreover, existing studies on the commercial green property and green building at large seem to be predominantly descriptive and qualitative and therefore lack rigorous quantitative empirical utility. There appear to be a few authors who have attempted to examine the correlations among the green building drivers. Investors and developers are not only interested in correlations but also which variables, such as monetary green tax incentives and available green skills, have the most causal effects on the nature of supply for green commercial building factors.

There is a lack of theoretical context and explanations due to the limited studies in this subject area. It is also hard to find studies in commercial green property supply that are based on structural equation modelling (SEM), which are popular in explaining causal relationships among constructs and variables as well as testing the reliability and validity of the model's instruments. In this study, Social Cognitive Theory (SCT) was used to provide context to the developers' motivational drivers for green building and explain structural relationships among the constituent constructs or factors. SEM was used to test the structural relationships in terms of causal effects as well as to test and validate the research instruments and model as a whole.

Thus, the objectives of this study are to (1) develop and validate a model of factors affecting the supply of green commercial property investment and (2) determine the causal effect of monetary green tax incentives and available green skills on life cost saving motivations, government policies and green certification, developers' expected rate of return motivations, and market strategy benefit motivations in relation to green building supply and investment.

It is hoped that the findings and model resulting from this study will have a strong empirical utility for researchers, developers, investors, and governments involved in green building who are seeking practicable explanations for significant empirical evidence of causal relationships between factors of green commercial property investment, green building incentives and green building skills.

Part IV Words and Phrases

Unit 1 Words and Phrases of Literatures

A

abandon	*v.*	放弃
abbreviation	*n.*	缩写
abrasion	*n.*	擦伤,磨损
abundance	*a.*	丰富,充裕,大量
access road		入口
accessibility	*n.*	可达性,可接近性
acid corrode		酸性腐蚀
acre	*n.*	英亩(= 6.07 亩)
acronym	*n.*	简称,只取首字母的缩写词
adapter	*n.*	适配器,转换接头,附件
admeasurement contract		计价合同
admixture	*n.*	掺合剂
adverse	*a.*	不利的
aerodynamic	*a.*	空气动力的
aggregate	*n.*	骨料,集料
agitation	*n.*	拌和
air conditioning		空气调节
air-entrained concrete		加气混凝土
akin	*a.*	类似的,同样的(常跟 to)
alcove	*n.*	凹室,壁龛
alkaline	*a.*	碱性(的)
allocate	*v.*	分配,分派,配给
allowable stress		允许应力
allowable stress approach		允许应力法
allowance	*n.*	留量,容差,补助
alkali-aggregate reaction		碱-骨料反应
alkalinity	*n.*	(强)碱性, 碱度
alternative	*n.*	比较方案
	a.	交替的,变更的, 比较的
aluminum (aluminium)	*n.*	铝
aluminum alloy		铝合金
amalgam	*n.*	混合物,软的混合物
amenity	*n.*	舒适,适宜,愉快

analogous	*a.*	类似的,相似的
analogous to		类似于,相似于
analytical	*a.*	分析的,解析的
analytical model		分析模型
ancillary	*a.*	辅助的,附属的
angle	*n.*	角度
	n.	支座
at right angle to		与……正交
anodic	*a.*	阳极的
anticipate	*v.*	预期
apartment	*n.*	公寓,住所
applied mechanics		应用力学
architect	*n.*	建筑师
architectural	*a.*	建筑(学)的
arena	*n.*	表演场
arguable	*a.*	可争辩的,可论证的
assessment	*n.*	估计,评价
available	*a.*	有效的

B

baffling	*a.*	阻碍……的,起阻碍作用的
balance	*n.*	平衡
bar	*n.*	法庭,律师的职业
bar examination		资格考试,职业考试
batch	*n.*	配料
bath pool		浴室
beam	*n.*	梁
bearing wall		承重墙
(be) liable for		对……负责的
be entitle to		给……权利
behavior	*n.*	状态
limit behavior		极限状态
bedroom	*n.*	卧室
bend	*n.*	弯曲
bending	*n.*	弯曲
Bessemer converter		贝式转炉
Bessemer process		贝色麦法
binder	*n.*	黏结料
binding agent		黏结料,结合料,黏合剂
binding capacity		黏结力
binding contract		有(法律)约束力的合同
bitumen	*n.*	沥青
blast	*n.*	爆破

blend	*n.*, *v.*	混合
bolt	*n.*	螺栓
bolting	*n.*	螺栓（连接）
borehole	*n.*	钻孔
breakage	*n.*	破坏
breach	*n.*, *v.*	破坏，违反
breach of contract		违约
brick of burnt clay		黏土烧结砖
bridge	*n.*	桥梁
brittle	*a.*	脆性的
brittle fracture		脆裂
buckling	*n.*	压曲，弯折
budget	*n.*, *v.*	预算，作预算，编入预算
building	*n.*	建筑，房屋
building line		建筑红线
bulk	*n.*	湿胀性
bulldozer	*n.*	推土机，开土机，压路机
by contrast		对照之下

C

cabinetwork	*n.*	细木工，细木家具
cable-stayed bridge		斜拉桥
cage-like	*a.*	骨架状的
calibration	*n.*	标度，校准
capability	*n.*	能力
capacity-reduction factor		承载能力折减系数
capillary suction		毛细作用
capital	*n.*	资金
carbon	*n.*	碳
carbon dioxide		二氧化碳 CO_2
carbon steel		碳素钢
carbonate	*n.*	碳酸盐
carbonate lime		碳酸钙 $CaCO_3$
carbonation	*n.*	碳化
cathodic	*a.*	阴极的
cavitations	*n.*	气穴现象
ceiling	*n.*	天棚
cellular	*a.*	细胞的，分格的，多孔状的
cement	*n.*	水泥
cement hydration		水泥水化作用
cement mortar		水泥浆
cement paste		水泥浆
centroid	*n.*	形心，矩心

chloride	*n.*	氯化物，漂白粉
chainlike	*a.*	链式的
church	*n.*	教堂
circumferential	*a.*	周围的，环形的，环绕的
circumstance	*n.*	情况
circumvent	*vt.*	回避，绕过，围绕
civil engineering		土木工程
cladding	*n.*	镀层，保护层
claim "damages"		索赔
clamp	*n.*，*v.*	夹子，夹钳；卡紧
clamshell	*n.*	抓斗，蛤壳式挖泥机
clarify	*v.*	澄清，阐明，净化，解释
classification	*n.*	分类
clay	*n.*	黏土
clear span		净跨
clear spacing		净距
clutter	*n.*	混乱，杂乱，干扰
	v.	弄乱，使混乱
coarse (fine)-grained soils		粗（细）粒土
cobble	*n.*	石子，圆石块
code	*n.*	规范，法规（则），（代）码
cohesion	*n.*	黏结力，内聚力
collapse	*v.*	倒塌
column space		柱距
combination	*n.*	组合
commercial concrete		商品混凝土
commission	*n.*	委托书
communication	*n.*	交通，联络
compact	*v.*	压实
compaction	*n.*	压实
compatibility	*n.*	兼容性，相容性，适用性，互换性
competent	*a.*	竞争的
comply	*vi.*	应允，遵照（常跟 with）
composite structures	*n.*	组合结构
compound	*n.*	化合物
compressibility	*n.*	压缩性
compression	*n.*	压力
compressive strength		抗压强度
computer-aided		计算机辅助
concave	*a.*，*n.*	凹的，凹面
conceal	*v.*	隐蔽，隐瞒，把……隐藏起来
concentration	*n.*	浓度

concentric	*a.*	同轴的
concrete	*n.*	混凝土
concrete block		混凝土砌块
concrete cover		混凝土保护层
condition	*n.*	条件
conduct	*n.*，*v.*	行为，操行；引导，管理，传导
conduit	*n.*	管道，导管，水道，水管
confusion	*n.*	混乱，杂乱
congest	*v.*	充满
be congested with		充满，布满
conifer	*n.*	针叶树，松柏类植物
conservation	*n.*	储备
consolidation	*n.*	渗压，加强
	n.	捣实
constant	*n.*	常数
construction	*n.*	施工，建设
construction expertise		施工专家
construction phase		施工阶段
construction site		施工现场
consultant	*n.*	顾问，咨询者
contamination	*n.*	玷污，污染，污染物
content	*n.*	容量
continuous structure		连续结构
continuum	*n.*	连续体
contract	*n.*	合同
contract terms	*n.*	合同条款
contractor	*n.*	承包商
contraction	*n.*	收缩
converter	*n.*	炼钢炉，吹风转炉
convex	*a.*，*n.*	凸的，凸面
convey	*v.*	搬运
copper	*n.*	铜
corridor	*n.*	走廊，通道，过道
corrosion	*n.*	腐蚀，侵蚀
corrugate	*v.*	弄皱，使起皱纹，
	a.	起皱的，起波纹的
cost-reimbursable contract		成本补偿合同
counteract	*v.*	抵抗
counterbalance	*v.*	抵消，平衡
crane	*n.*	起重机
credibility	*n.*	可接受性
creep	*n.*	徐变

criteria *n.* 标准
criterion *n.* 规则；标准
critical *a.* 临界的
critical section 临界截面
crushed rock 碎岩
cure *v.* 养护
curriculum *n.* 课程，学习计划
curtain wall 悬墙，幕墙，围护墙
custom-designed 惯例
cylindrical *a.* 圆柱体的

D

dam *n.* 坝
gravity bam 重力坝
earth-fill embankment bar 填土筑堤坝
damp down 减轻，减缓
damper *n.* 阻尼（减速，减振）器
darby *n.* 刮尺
dead load 恒载
decay *v.*，*n.* 腐朽（化），衰变
deck *n.* 甲板，舱面，桥面，层面
defect *n.* 缺陷
defect liability period 缺陷责任期
deflection *n.* 挠曲
deformable body 变形体
deformation *n.* 变形
deicing salt 除冰盐
delay *n* 延误
deleterious *a.* 有害的
deposit *v.* 沉淀
depression *n.* 降低，不景气，萧条期
depth *n.* （截面）高度，深度
derrick *n.* 井架
design *n.* 设计
destruction *n.* 破坏
destructive *a.* 有害的
deterioration *n.* 恶化，退化
deterrent *a.* 制止的，威慑的，
n. 制止物，威慑因素
dewater *n.* 降水
diffusion *n.* 扩散
digitize *v.* 数字化，将资料数字化
dimension *n.*，*v.* 尺寸，尺度；定尺寸
dining room 餐厅

dismantle	*v.*	拆除，拆卸，粉碎
disposal	*n.*	控制权
dissipative	*a.*	消（分，扩）散的，散逸的
dissolution	*n.*	分解
dome	*n.*	圆顶
domestic building		民用建筑
Douglas fir	*n.*	美国松
dragline	*n.*	拉索，拉铲挖土机
duct	*n.*	管道，通道，预应力筋孔道
ductility	*n.*	韧性
dump sand		湿砂
duplicate	*v.*	复写，复制，转录，重复
durability	*n.*	延性
durable	*a.*	耐久的
duration	*n.*	持续时间
dusting	*n.*	起尘
dwelling	*n.*	居住
dynamic	*a.*	动力的

E

earth-fill embankment dam		填土坝
eccentric	*a.*	偏心的
elastic modulus	*n.*	弹性模量
elastically	*ad.*	弹性的
elasticity	*n.*	弹性，弹力，弹性力学
elastic-plastic displacement		弹塑性位移
electrical and mechanical system		机电系统
electrolytic	*n.*	电解
element	*n.*	构件
elementary	*a.*	基本的，初步的
elongation	*n.*	伸长，延长
embed	*v.*	埋入
employer	*n.*	业主
employment	*n.*	办公，写字楼
encounter	*v.*，*n.*	遭遇，遇到，相遇
engineer	*n.*	工程师
	n.	发动机
engineering	*n.*	工程
engineering graduate		工科毕业生
engineer representative		工程师代表
environmental or sanitary engineering		环境卫生工程
environmental planning		环境设计，环境规划
equivalent	*a.*	等效的
evaluation	*n.*	评估，评价
evaporate	*v.*	蒸发

evidence	*n.*	证据
excavation	*n.*	挖掘，挖方，发掘
execute	*v.*	实施
execution	*n.*	执行，完成，实施，施工
exert	*v.*	施工
exhibit	*v.*	显示，呈现
exhibition hall		展览厅
existing structures		既有结构
expansion	*n.*	伸长
expertise	*n.*	专门技能，专门知识
explicitly	*ad.*	明晰地，明确地
exploration	*n.*	勘察
expulsion	*n.*	排除
exterior	*a.*	外部的，外面的，外部，表面

F

fabric	*n.*	构造，结构
factor	*n.*	因素
safety factor（factor of safety）		安全因素
facility	*n.*	设备，实验室
failure	*n.*	破坏
fatigue	*n.*	疲劳
feasibility study		可行性研究
fiber	*n.*	纤维
field（mode）test		现场（模型）试验
fireproof	*a.*	耐火的，防火的
flange	*n.*	（梁的）翼缘
flexible-shaft vibrator		柔轴振捣器
flexural crack		挠曲裂缝
float	*v.*	持平
floor area		占地面积
foreseen	*v.*	预见
formality	*n.*	正式手续
formwork	*n.*	模板，支模
foundation	*n.*	基础
fracture	*n.*	断裂
fragment	*n.*	断片，碎块，凝固
frame	*n.*	框架
framework	*n.*	构架，框架，结构
frequency	*n.*	频率
freeze/thaw cycle		冻融循环
frictional resistance		摩擦阻力
fulfillment	*n.*	履行

G

gasoline	*a*.	燃气的
geodetic	*n*.	大地测量学
geodetic surveying		大地测量学
geological origin		地质成因
girder	*n*.	（桥梁或建筑物的）大梁，桁架
gorge	*n*.	峡谷
governing condition		控制条件
government jurisdiction		政府行政区
grain	*n*.	纹理
gravel	*n*.	砾石
grout	*v*.	灌浆
gross	*a*.，*n*.	总的，显著的，总额
ground level	*n*.	地面水平线
gymnasium	*n*.	体育馆，健身房
grain	*n*.	颗粒，纹理，粒面
granular	*a*.	颗粒状的
graphic	*a*.	图示的
gravel	*n*.	砾石
gravitational force		重力
gust	*n*.	（一）阵（骤）风（雨）
gymnasium	*n*.	体育馆

H

habitation	*n*.	居住，住所
hand-held	*a*.	手持的
hardcopy	*v*.，*n*.	硬拷贝
hardware	*n*.	硬件，部件
haul	*v*.	拖，拉，用力拖拉，拖运
height-to-width ratio		高宽比
high-alkali	*a*.	高碱性的
hoop	*n*.，*v*.	箍筋；箍住
horizontal axis		水平轴
hospital	*n*.	医院
hotel	*n*.	酒店
hugging stress		握裹（应）力
hydrate	*v*.	水化
hydration	*n*.	水化（作用）
hydraulic	*a*.	液压的，水硬性的
hydraulics	*n*.	水利学
hydrostatic	*a*.	液体静力（学）的
hydroxyl	*a*.	氧化的
hyperbola	*n*.	双曲线

hyperbolic	*a.*	双曲线的，夸大的

I

ignorance	*n.*	未知数
immersion-type vibrator		沉入型振导器
impact	*n.*	冲击力
impermeable	*a.*	不能渗透的
impose	*v.*	施加
incipient	*a.*	早期的
inclined-axis	*n.*	斜轴
incompressible	*a.*	不可压缩的
industrial (civil) buildings		工业（民用）建筑
ineffective	*a.*	无效的
inert	*a.*	惰性的，迟钝的
inflation	*n.*	通货膨胀、通胀率、充气
ingenuity	*n.*	独创性，机灵
ingredient	*n.*	（混合物）成分，配料，组分
insertion	*n.*	插入，嵌入，插页
insula	*n.*	群房，公寓
internal (gross) force		（总）内力
internal vibrator		内部振导器
intensity	*n.*	集度，强度，密（集）度
intensity of force		力的集度（应力）
internal (gross) force		（总）内力
interim	*a.*	间歇的，暂时的
interim payment (certificates)		中间付款（验收，或证书）
intuition	*n.*	直觉，直观
intuitively	*ad.*	直觉上，直观上
investment	*n.*	投资
invitation to tender		邀标
invoice	*n.*	发票，发货单，开发票，记清单
involve	*v.*	包括，包含
ion	*n.*	离子
ipso-facto		照那个事实，根据事实本身
iron rod		铁条（棒）
irrecoverable	*a.*	不能恢复的
issue	*n.*	发布

J

joystick	*n.*	控制杆，操纵杆，游戏杆
jurisdiction	*n.*	管辖权，权限
justification	*n.*	正当理由，认为正当

K

keyboard	*n*.	键盘，用键盘写入
kiln	*n*.	窑
kip	*n*.	千磅（重量单位）
kitchen	*n*.	厨房

L

larder	*n*.	食品室，储藏室
lateral	*n*.	侧面的
lateral motion		侧移
lateral stability		侧向稳定
lateral sway		倾斜，水平摆动
launching pads		发射台
layout	*n*.	计划，方案，布局，格式
legal	*a*.	法律的
liberal	*a*.	开朗的，（思想）自由开朗的
liberalize	*v*.	放宽（限制）范围，使自由化
lifetime	*n*.	寿命
lighting system		照明系统
lime	*n*.	石灰
lime fraction		氢氧化钙 $Ca(OH)_2$
limestone	*n*.	石灰石
limit state		极限状态
linear (circular) prestressing		线（环）形预应力
lintel	*n*.	楣，（门窗）过梁
live load		活载
living room		起居室
load	*n*.	荷载
dead load		恒载
live load		活载
load effect		荷载效应
longitudinal	*a*.	纵向的
long-time	*a*.	长期的
lumped	*a*.	整块的
lump sum contract		总价合同

M

mainframe	*n*.	主机，大型机
maintenance	*n*.	维修
maintenance period expired		维修期满
mandatory	*a*.	必须遵循的，命令的
manganese	*n*.	（化学元素）锰
manufacture	*n*.	加工，制造

manipulate	*v.*	操纵，操作，生成
margin	*n.*	空白，边缘，极限，富余
margin of safety (factor of safety)		安全系数
magnitude	*n.*	大小，量，量值
masonry	*n.*	圬工，砌筑
master	*a.*	主要的，总的，熟练的
mechanics	*n.*	地质学专家
mechanism	*n.*	机理（制），技巧，手法
membrance-forming curing compounds		薄膜养护剂
mercantile	*a.*	商业的，贸易的
metallurgical	*a.*	冶金的
metropolitan	*a.*	大城市的
midspan	*n.*	跨中
mileage	*n.*	利润
mill-rolled	*n.*	轧钢机
minor	*a.*	不严重的
mixer	*n.*	搅拌机
mix	*v.*	搅拌
monitor	*n.*	监视器，监督程序
monochrome	*n.*	单色，单色图像
monolithic	*a.*	（建筑等）庞大而无特点的，巨大而单调的
mortar	*n.*	砂浆，灰浆
motion	*n.*	运（移）动，运行（转）机械
multi-story	*a.*	多层的

N

nature	*n.*	性质
natural lighting		自然采光
natural water table		自然水位
negligible	*a.*	可以忽略的，微不足道的
negotiation	*n.*	协商
neutral plane		中性面
non-cohesive	*a.*	无黏结的
nondimensional	*a.*	无尺寸的，无单位的
non-elastic	*a.*	非弹性的
nonlinear	*a.*	非线性的
nylon	*n.*	尼龙
nuisance	*n.*	麻烦事，妨害行为

O

obligation	*n.*	义务，责任
observe	*v.*	遵守
offer	*n.*	报价
on-the-job		在现场的，在职的

opaque	*a.*	透明的，不透光的，不透明体
organic plastic		有机塑料
orient	*n.*	定向，定位
orientation	*n.*	方位，方向
overhead	*n.*	管理费
	n.	超载
overlap	*n.*	重叠，搭接
overrun	*v.*	超过，超出，超限
overturning	*n.*	倾覆

P

paddle	*n.*	浆状
pad or isolated foundation		独立基础
parabolic	*a.*，*n.*	抛物面（线，体）的
parallel	*v.*	平行
parameter	*n.*	因素
park	*n.*	停车场
partial prestressing		部分预应力
particle size		粒径
partition	*n.*	隔墙
passivation	*n.*	钝化作用
passive layer		钝化层
paste	*n.*	胶结材料
patent	*n.*	首创，专利
pattern	*n.*	形式
pay for		支付
payroll	*n.*	薪水册，发放工资额，工资单
pedestrian	*n.*	行人，步行者
penetrate	*v.*	渗透
penetration	*n.*	贯入
performance	*n.*	任务，性能
perimeter	*n.*	周长，环行
permeability	*n.*	渗透，渗透性
permanent	*a.*	永久的，恒久的
permanent stress		永久应力
permissible	*a.*	容许的
petrography	*a.*	岩石（学）的，岩相（学）的
philosophy	*n.*	（基本）原理，哲学，宗旨
pile foundation		桩基础
placement	*n.*	浇注
plan	*n.*，*v.*	设计，计划
plastic	*n.*	塑料

plaza *n.* 广场，集市场所，大空地
plotter *n.* 绘图机，绘图仪，图形显示器
plumbing system 卫生设备系统
plunge *v.* 插入
pneumatic-tired roller 气动碾压机
poker vibrator 插入式振导器
point of collapse (collapse point) 破坏点
polymer *n.* 聚合物
porous *a.* 多孔性的，能渗透的
popouts *v.* (混凝土表面) 剥落
porosity *n.* 多孔性
port *n.* 断口，通讯口，进出口
portal frame 门式框架
Portland cement 波特兰水泥，硅酸盐水泥
positive *a.* 正的
post *n.* 柱，支柱
postpone *v.* 拖延，延迟
pour *v.* 浇注
power house 发电站
pozzolan *n.* 火山灰
pragmatic *a.* 重实效的，实际的
precast unit 预制构件
precondition *n.* 前提
preload *v.* 预加荷载
preliminary *a.* 预备的，初步的
n. 前期
preliminary planning 初步规划
premature *a.* 过早的
premium *n.* 奖金，额外费用，高级，保险费
a. 质量改进的，特级的，高昂的，优质的
prestressed concrete 预应力混凝土
prestressing force 预应力
pressimistic load 不利荷载
pressure *n.* 压力
pretensioning (posttensing) concrete 先（后）张法
price-based 价格基础
prismatic *a.* 棱柱（形）的
prismatic bar 等直杆
probability *n.* 概率论，可能性
processor *n.* 处理机，处理器，处理程序
productivity *n.* 生产力，劳动生产率

profit	*n.*	利润
project	*n.*	项目
projection	*n.*	规划
promoter	*n.*	发包者
prompt	*n.*, *v.*	提示符，提示
proof test		验证试验
proportion	*n.*	尺寸，面积
provision	*n.*	条款
public buildings		公共建筑
pulverize	*v.*	研磨
pump	*v.*	泵灌
pyramid	*n.*	金字塔，四面体

Q

quadrilateral	*a.*, *n.*	四边（角）形，四方面的
quantify	*v.*	确定数量，用数量表示
quarter	*n.*	住处
quasi-global	*a.*	准（拟，半）总体的

R

rammer	*n.*	夯实机
Random Access Memory		随机存取存储器
raft foundation		筏基
reactant	*n.*	反应物
reactive aggregate		活性骨料
reactive force		反力
read-mixed concrete		预拌混凝土，商品混凝土
realistic	*a.*	现实（主义的）
rebind	*v.*	重捆，重新装（包扎）
rebound	*v.*	弹回，回跳
recruit	*v.*	招聘
redraw	*v.*	重画，刷新屏幕
reinforcement	*n.*	加强，强化
rehabilitation	*n.*	复原，更新
reimburse	*v.*	补偿
reinforced bar	*n.*	配筋
reinforced concrete		钢筋混凝土
remove	*v.*	拆除
remove formwork		拆除模板
resin	*a.*	树脂，胶质，人造树脂
resistance	*n.*	抗力
response	*n.*	应（回）答，反（响）应曲线，特性曲线
restraint	*n.*	约束

resultant	*n.*	合力
retention money		保留（滞留金）
rib	*n.*	肋
ridge	*n.*	脊，岭
rigid frame		刚性框架
rivet	*v.*	铆接
roadway	*n.*	车行道，路面
rocket storage facilities		火箭库
rotating or turning moment		旋转力矩，扭转力矩
rubber band		橡胶带（条）
rugged	*a.*	崎岖的，艰难的
rupture	*n.*	破裂，断裂、爆裂
rust	*v.*	生锈

S

safety factor		安全系数
sand	*n.*	砂子
saturate	*v.*	浸透，渗透，使充满
scaling	*v.*	分层
schedule	*v.*	将……列表，安排
schedule production		钢筋表，计划表
science-orient		注重科学的
scientific publication		科学刊物
scraper	*n.*	铲运机，刮土机，平土机
screed	*v.*	找平
seashell	*n.*	海贝，贝壳
secure	*v.*	保证
section	*n.*	截面
secular	*a.*	世俗的，现世的，非宗教的
segmental	*a.*	弓形的，拱形的
segregate	*v.*	离析
segregation	*n.*	分层，离析
sensitive	*a.*	敏感的
service (live, dead) load		使用（活，静）荷载
service performance		使用性能
serviceability	*n.*	适用（性），耐用（性）
set	*v.*	凝固
set out		制定
settlement	*n.*	沉陷，沉降
sewer	*n.*	下水道
shaft	*n.*	轴
shear	*n.*, *v.*	剪切，剪力

shear strength		抗剪强度
shed	n.	小棚，小屋
sheen	n.	光泽
sheet	n.	（薄）板，杆
shovel	n.	铲，挖掘机，单斗挖土机
shrinkage	n.	收缩
sideway	n.	侧面
silicon	n.	硅，硅元素
silt	n.	淤泥，残渣
silo	n.	筒仓，竖井，（导弹）发射井
simultaneously	a.	同时的
sitting room		起居室
skeleton	n.	骨架
slab	n.	平板
slate	n.	板岩，石板瓦，铺石板
slenderness	n.	细长（度）
slender proportion		长细比
slip form	n.	滑动模板，滑模（施工法）
slump	n.	坍落度
socioeconomic	a.	社会经济（学）的
soil investigation		土质勘测
soil mechanics		土力学
soil stabilization		土壤稳定
solidify	v.	固化，固结，凝固
soundest	a.	安全的，坚固的
space	n.	空间，面积，跨度
spalling	n.	剥落，脱落，散裂，剥离
spare-time activities		业余活动
specification	n.	（常pl.）规范，详述，规格，说明书
spherical	a.	球（形）的
sports arenas		运动场
spray	v.	喷，喷射
spread footing		扩展基础
spruce	n.	云杉，云杉木
square-root-time	n.	时间的平方根
stable	a.	稳定的
stack	v.	堆叠，成堆，整齐地堆起
stanchion	n.	（用以支撑的）杆，支柱
station	n.	车站
stationary	a.	坚固的
steel	n.	钢，钢材
steel rods		钢筋，钢条

stiffen	*v.*	硬化
stiffness	*n.*	劲度，刚度，硬度
storage	*n.*	存储器
story	*n.*	层
strain	*n.*	应变
stratification	*n.*	分层，层理
stretch	*v.*	伸展
strength	*n.*	强度
strength design		强度设计
great strength		高强度，高强
stress	*n.*	应力
stress intensity		应力强度
strip foundation		条形基础
structure	*n.*	结构
structural materials		建筑材料
structural system		结构系统
study	*n.*	书房
stylus	*n.*	唱针，铁壁（汉字输入用），指示笔
subsoil	*n.*	下（亚）层土，地基下层土
subsidiary	*a.*	辅助的，补充的，次要的
substantial	*a.*	大量的，价值巨大的，重大的，大而坚固的，结实的，牢固的
substructure	*n.*	下部结构
sulfate	*n.*	硫酸盐（酯）
sulfate-resistant	*a.*	耐硫酸盐的
summit	*n.*	顶点
superficial	*a.*	表面的，肤浅的
superstructure	*n.*	上部结构
supervision	*n.*	管理，监控
surpass	*v.*	超过，胜过
supply	*v.*	供应
support	*n.*	支撑，支座
swelling	*n.*	膨胀
synthetic	*a.*	合成的，人造的，人造树脂

T

tamp	*v.*	夯实
tar-like	*a.*	焦油般的
target cost contract		目标成本合同
tender	*n.*，*v.*	标书；招投标
telephone exchange		电话交换室
tenement	*n.*	出租的房子，经济公寓
tensile strength		抗拉强度

tensile yielding		抗拉屈服
tension	*n.*	拉力
terrain	*n.*	地域，地带，领域
textile industry		纺织工业
thaw	*v.*	融化，解冻
the fire hazard		火灾
the groundwater level		地下水位
the life of the structure		结构寿命
the lines of stress		应力线
the passenger elevator		载客电梯
theatre	*n.*	戏院
thermoplastic	*a.*	热塑性的
thermosetting	*a.*	热凝性的，热固性的
thrust	*n.*，*v.*	推，推力，轴力
tie rod		连杆
tile	*n.*	瓷砖，瓦，花砖
timber	*n.*	木材，木料，原木
timber work		木结构
time-dependent		与时间有关的，时变的，时效的
toilet	*n.*	卫生间
topographic（al）	*a.*	地形学（的）
topography	*n.*	地形（势，貌）
torsional capacity		扭力
town-planning	*n.*	城镇规划
tractor	*n.*	牵引机
trail and error		反复试验，试错法，尝试法
transient	*a.*，*n.*	瞬间，瞬态
transparent	*a.*	透明的
transportation	*n.*	运输
transverse	*a.*	横向的
trial pits		试验坑
tricalcium aluminate		三钙铝酸盐
tropical	*n.*，*a.*	回归线（*pl.*）热带；热带的
truck mixing concrete		车载搅拌混凝土
truss	*n.*	桁架
turnbuckle	*n.*	松紧螺旋扣
twist	*v.*，*n.*	扭转，编织
twisting moment		扭矩

U

ultimate strength		极限强度
utilize	*v.*	安装
ultimate limit state		极限状态

ultimate tensile strength		极限抗拉强度
unbonded	*a*.	无黏结的
unconservation	*n*.	无储备
underground	*a*.	地下的
unduly	*ad*.	过度地，不适当地
unequal	*a*.	不均匀的
unequivocal	*a*.	不含糊的，明确的
uneven	*n*.	不均匀的
uniform	*a*.	均匀的
unit area		单位面积
unload	*v*.	卸载
unjust	*a*.	不正当的
unsatisfactory	*a*.	不良的
ultimate	*a*.	极限的
urbanization	*n*.	都市化，都市集中化

V

variable	*n*.	变量
variety	*n*.	变更
vellum	*n*.	精制犊皮纸，牛皮纸
ventilate	*v*.	使通风，使通气 给……装置通风设备
versatile	*a*.	多用途的，多方面适应的
vertical plane		竖向平面
via	*v*.	经过
vibrating screed		振捣板
vibration	*v*.	振捣
vibratory-type roller		振捣型碾压机
vicinity	*n*.	附近，接近
video	*n*.	试品，影响，电视（图像）
vis-à-vis	*ad*. *prep*.	相对 与……相比较
void	*n*.	孔隙，孔隙率
volcanic ash		火山灰
vulnerable	*a*.	易受伤害的，有弱点的

W

water pressure		水压力
water supply systems		供水系统
watertightness	*n*.	不漏水
wear resistant		抗磨
web	*n*.	（梁的）腹板

weld	*v.*	焊接
wellpoint	*n.*	降低地下水位的井点，深坑点
wind force		风力
wind tunnel (test)		风洞（试验）
wire stands		钢绞线
withdrawn	*v.*	起草
withhold	*v.*	扣留
workability	*n.*	和易性
workmanship	*n.*	工艺，技巧

Y

yield point		屈服点
yield strength		屈服强度

Z

zinc	*n.*	锌
zoning	*n.*	分区，区域化，分地带

Unit 2 Words and Phrases of Engineering Codes

2.1 Load 荷载

permanent load	永久荷载
variable load	可变荷载
accidental load	偶然荷载
representative values of a load	荷载代表值
design reference period	设计基准期
characteristic value/nominal value	标准值
combination value	组合值
frequent value	频遇值
quasi-permanent value	准永久值
design value of a load	荷载设计值
load effect	荷载效应
load combination	荷载组合
fundamental combination	基本组合
accidental combination	偶然组合
characteristic/nominal combination	标准组合
frequent combinations	频遇组合
quasi-permanent combinations	准永久组合
equivalent uniform live load	等效均布荷载
tributary area	从属面积
dynamic coefficient	动力系数
reference snow pressure	基本雪压
reference wind pressure	基本风压
terrain roughness	地面粗糙度

2.2 Subgrade and Foundation 地基基础

subgrade，foundation soil	地基
foundation	基础
characteristic value of subgrade bearing capacity	地基承载力特征值
gravity density，unit weight	重力密度（重度）
rock discontinuity structural plane	岩体结构面
standard frost penetration	标准冻深
allowable subsoil deformation	地基变形允许值
soil-rock composite subgrade	土岩组合地基
ground treatment	地基处理
composite subgrade，composite foundation	复合地基
spread foundation	扩展基础
non-reinforced spread foundation	无筋扩展基础
pile foundation	桩基础

retaining structure	支挡结构

2.3 Timber Structure 木结构

timber structure	木结构
log	原木
sawn lumber	锯材
square timber	方木
plank	板材
dimension lumber	规格材
glued lumber	胶合材
moisture content of wood	木材含水率
parallel to grain	顺纹
perpendicular to grain	横纹
at an angle to grain	斜纹
glued laminated timber (Glulam)	层板胶合木
swan and round timber structures	普通木结构
light wood frame construction	轻型木结构
stud	墙骨柱
visually stress-graded lumber	木材目测分级
machine stress-rated lumber	木材机械分级
truss plate	齿板
wood-based structural-use panels	木基结构板材
shear wall of light wood frame construction	轻型木结构的剪力墙

2.4 Concrete Structure 混凝土结构

concrete structure	混凝土结构
plain concrete structure	素混凝土结构
reinforced concrete structure	钢筋混凝土结构
prestressed concrete structure	预应力混凝土结构
pretensional prestressed concrete structure	先张法预应力混凝土结构
post-tensioned prestressed concrete structure	后张法预应力混凝土结构
cast-in-situ concrete structure	现浇混凝土结构
prefabricated concrete structure	装配式混凝土结构
assembled monolithic concrete structure	装配整体式混凝土结构
frame structure	框架结构
shearwall structure	剪力墙结构
frame-shearwall structure	框架-剪力墙结构
deep flexural member	深受弯构件
deep beam	深梁
ordinary steel bar	普通钢筋
prestressing tendon	预应力钢筋
degree of reliability	可靠度
safety class	安全等级
design working life	设计使用年限
load effect	荷载效应
load effect combination	荷载效应组合

fundamental combination 基本组合
characteristic combination 标准组合
quasi-permanent combination 准永久组合

2.5 Steel Structure 钢结构

strength 强度
load-carrying capacity 承载能力
brittle fracture 脆断
characteristic value of strength 强度标准值
design value of strength 强度设计值
first order elastic analysis 一阶弹性分析
second order elastic analysis 二阶弹性分析
buckling 屈服
post-buckling strength of web plate 腹板屈服后强度
normalized web slenderness 通用高厚比
overall stability 整体稳定
effective width 有效宽度
effective width factor 有效宽度系数
effective length 计算长度
slenderness ratio 长细比
equivalent slenderness ratio 换算长细比
nodal bracing force 支撑力
unbraced frame 无支撑纯框架
frame braced with strong bracing system 强支撑框架
frame braced with weak bracing system 弱支撑框架
leaning column 摇摆柱
panel zone of column web 柱腹板节点域
spherical steel bearing 球形钢支座
couposite rubber and steel support 橡胶支座
chord member 主管
bracing member 支管
gap joint 间隙节点
overlap joint 搭接节点
uniplanar joint 平面管节点
multiplanar joint 空间管节点
built-up member 组合构件
composite steel and concrete beam 钢与混凝土组合梁

2.6 Masonry Structure 砌体结构

masonry structure 砌体结构
reinforced masonry structure 配筋砌体结构
reinforced concrete masonry shear wall structure 配筋砌块砌体剪力墙结构
fired common brick 烧结普通砖
fired perforated brick 烧结多孔砖
autoclaved sand-lime brick 蒸压灰砂砖

autoclaved flyash-lime brick 蒸压粉煤灰砖
concrete small hollow block 混凝土小型空心砌块
morter for concrete small hollow 混凝土砌块砌筑砂浆
grout for concrete small hollow 混凝土砌块灌孔混凝土
plastered wall 带壁柱墙
rigid transverse wall 刚性横墙
cavity wall filled with insulation 夹心墙
structural concrete column 混凝土构造柱
ring beam 圈梁
wall beam 墙梁
cantilever beam 挑梁
design working life 设计使用年限
static analysis scheme of building 房屋静力计算方案
rigid analysis scheme 刚性方案
rigid-elastic analysis scheme 刚弹性方案
elastic analysis scheme 弹性方案
upper flexible and lower rigid complex multistorey building 上柔下刚多层房屋
types of roof or floor structure 屋盖、楼盖类别
ratio of hight to sectional thickness of wall or column 砌体墙、柱高厚比
effective support length of beam end 梁端有效支承长度
calculating overturning point 计算倾覆点
expansion and contraction joint 伸缩缝
control joint 控制缝
category of construction quality control 施工质量控制等级

2.7 Tall Building 高层建筑

tall building 高层建筑
building height 房屋高度
frame structure 框架结构
shearwall structure 剪力墙结构
frame-shearwall structure 框架-剪力墙结构
slab-column shearwall structure 板柱-剪力墙结构
tube structure 筒体结构
frame-corewall structure 框架-核心筒结构
tube in tube structure 筒中筒结构
mixed structure，hybrid structure 混合结构
transfer member 转换结构构件
transfer story 转换层
story with outriggers and/or belt member 加强层

2.8 Seismic of Structure 结构抗震

seismic fortification intensity 抗震设防烈度
seismic fortification criterion 抗震设防标准
earthquake action 地震作用

design parameters of ground motion 设计地震动参数
design basic acceleration of ground motion 设计基本地震加速度
design characteristic period of ground motion 设计特征周期
site 场地
seismic concept design of buildings 建筑抗震概念设计
seismic fortification measures 抗震措施
details of seismic design 抗震构造措施

2.9 Building Engineering 建筑施工

building engineering 建筑施工
quality of building engineering 建筑工程质量
acceptance 验收
site acceptance 进场验收
inspection lot 检验批
inspection 检验
evidential testing 见证取样检测
handing over inspection 交接检验
dominant item 主控项目
general item 一般项目
sampling inspection 抽样检验
sampling scheme 抽样方案
counting inspection 计数检验
quantitative inspection 计量检验
quality of appearance 观感质量
repair 返修
rework 返工
control grade of construction quality 施工质量控制等级
type inspection 型式检验
continuous seam 通缝
supposititious seam 假缝
reinforced masonry 配筋砌体
core column 芯柱
inspection at original space 原位检测
cast-in-situ concrete structure 现浇结构
prefabricated concrete structure 装配式结构
defect 缺陷
serious defect 严重缺陷
common defect 一般缺陷
construction joint 施工缝
inspection of structural performance 结构性能检验
part 零件
component 部件
element 构件
the smallest assembled rigid unit 小拼单元
intermediate assembled structure 中拼单元
set of high strength bolt 高强度螺栓连接副

slip coefficient of faying surface　抗滑移系数（螺栓连接）
test assembling　预拼装
space rigid unit　空间刚度单元
stud welding　焊钉（栓钉）焊接
ambient temperature　环境温度
swan and round timber structures　方木和原木结构
step joints　齿连接
structural glued-laminated timber　胶合木结构
glued-laminated timber (Glulam)　层板胶合木
Finger joints　指形接头
dimension lumber　规格材（木材）
lamination　层板
wood-frame construction　轻型木结构
studs　墙骨
joists　搁栅
structural wood-based panel　木基结构板材
wood-based panel　木基复合板材
structural plywood　结构胶合板
oriented strand board (OSB)　定向木片板
structural composite lumber (SCL)　结构复合木材
laminated veneer lumber (LVL)　旋切板胶合木
parallel strand lumber (PSL)　平行木片胶合木
laminated strand lumber (LSL)　层叠木片胶合木
prefabricated wood I-Joist　预制工字形木搁栅
truss plate　齿板
wood preservative　木材防护剂
retention　保持量
penetration　贯入度

2.10 Ground Treatment　地基处理

ground treatment　基础处理
composite subgrade, composite foundation　复合地基
characteristic value of subgrade bearing capacity　地基承载力特征值
cushion　换填垫层法
preloading　预压法
vacuum preloading　真空预压法
dynamic compaction, dynamic consolidation　强夯法
dynamic replacement　强夯置换法
vibroflotation, vibro-replacement　振冲法
sand-gravel pile　砂石桩法
cement-flyash-gravel pile　水泥粉煤灰碎石桩法
rammed soil-cement pile　夯实水泥土桩法
cement deep mixing　水泥土搅拌法
dry jet mixing　粉体喷搅法
deep mixing　深层搅拌法
jet grouting　高压喷射注浆法

lime pile 石灰桩法
lime soil pile 灰土挤密桩法
earth pile 土挤密桩法
piles thrusted-expanded in column-hammer 柱锤冲扩桩法
silicification grouting 单液硅化法
soda solution grouting 碱液法

2.11 Subgrade and Foundation and Construction of Water proof 地基基础及防水施工

geosynthetics foundation 土工合成材料地基
heavy tamping foundation 重锤夯实地基
dynamic consolidation foundation 强夯地基
grouting foundation 注浆地基
preloading foundation 预压地基
jet grouting foundation 高压喷射注浆地基
soil-cement mixed pile foundation 水泥土搅拌桩地基
soil-lime compacted column 土与灰土挤密桩地基
cement flyash gravel pile 水泥粉煤灰、碎石桩
pressed pile by anchor rod 锚杆静压桩
underground waterproof engineering 地下防水工程
grade of waterproof 防水等级
rigid waterproof layer 刚性防水层
flexible waterproof layer 柔性防水层
primary linning 初期支护
shield tunneling method 盾构法隧道
geosynthetics 土工合成材料

2.12 Construction of Ground and Roof and Decoration 地面、屋面及装饰装修施工

building ground 建筑地面
surface course 面层
combined course 结合层
base course 基层
filler course 填充层
isolating course 隔离层
troweling course 找平层
under layer 垫层
foundation earth layer 基土
shrinkage crack 缩缝
stretching crack 伸缝
lengthwise shrinkage crack 纵向缩缝
crosswise shrinkage crack 横向缩缝
life of water proof layer 防水层合理使用年限
a separate waterproof barroer 一道防水设防
dividing joint 分格法
full adhibiting method 满粘法
border adhibiting method 空铺法
spot adhibiting method 点粘法

strip adhibiting method	条粘法
cold adhibiting method	冷粘法
heat fusion method	热熔法
self-adhibiting method	自粘法
hot air welding method	热风焊接法
inversion type roof	倒置式屋面
elevated overhead roof	架空屋面
impounded roof	蓄水屋面
planted roof	种植屋面
building decoration	建筑装饰装修
primary structure	基体
detail	细部

Appendix

Appendix Ⅰ Common Affix in English for Special Purpose 专业英语常用词缀

常用前缀

内涵	词缀	意义	词例
否定	dis-，in-，non-，un-	不、无、非、未等	disorder 无序，inelastic 非弹性的，unloaded 未加载的，uncertainty 不定性
	mal-，mis-	不善，坏	malfunction 故障，miscalculate 算错
	de-，dis-，un-	去，解，消，除	decentralize 分散，disconnect 分离，unloading 卸载
	anti-，contra-，counter-	反、逆、对、抗	antirusting 防锈的，contraflexure 反向挠曲 counterbalance 抗衡
空间位置和方向	extra-	外，向外	extraneous 外加的，extrapolate 外推
	in-	向，向内	incurve 内弯，inclination 倾斜
	infra-	在下，在下部	infrastructure 基础，基础设施
	inter-	在……间，相互	interrelate 相互有关，interdepend 互相依赖
	intra-	在内，内部	intramural 城市内的，intranet 局域局
	mid-	中，间	midposition 中间位置，midsection 中间截面
	out-	外，向外，出	outline 轮廓，outward 向外的
	over-	在上面，在外	overground 地上的，overlook 俯视
	pre-，pro-	向前，在前	preface 序言，proceed 前进
	sub-，under-	下，在下面	subway 地下铁道，underground 地下的
	super-，sur-	在……上	superstructure 上部结构，surface 表面
时间次序	fore-	预先，先前	foreshock 前震，forecast 预报
	post-	后、次	posttensioned 后张的，postgraduate 研究生
	pre-	事先	prestress 预应力，precaution 预防
	re-	再，重新	renew 更新，rejustment 再调整
比较程度	extra-	格外，超越	extraordinary 非常的，extra-light 特轻的
	hyper-	超过，极度	hypersonic 超声的，hyperplane 超平面
	over-	超过，过度	overload 超载，overmix 拌和过度
共同相等	co-	共同，和	coexist 共存，cooperation 合作
	equi-	同等	equilibrium 平衡，equivalent 等价的
	sym-，syn-	同，共	symmetry 对称，synchronous 同步的

续表

内涵	词缀	意义	词例
通过遍及	dia-	通过,横过	diameter 直径,diagonal 对角线
	trans-	横过,贯通	transport 运输,transparent 透明的
数量	deca-, deci-	十,十分之一,1/10	decameter 米,decigram 分克
	hecto-, centi-	百,百分之一,1/100	hectoliter 公升,centimeter 厘米
	kilo-, milli	千,千分之一,1/1000	kilogram 千克,millimeter 毫米
	mega-, micro-	兆,微(百万分之一)	megacycle 兆周,microampere 微安培
	multi-	许多,多数	multimeter 万用表,multilateral 多边的
	hemi-, semi-	半,一半	hemicycle 半圆,semiconductor 半导体
其他	macro-, magni-	长,大,宏大,巨大	macroseism 强震,magnification 放大
	micro-	微小,小型	microphone 显微镜,microwave 微波
	ortho-	直,正,垂直	orthogon 矩形,orthograph 正视图
名词	-er	……者(人或物)	observer 观察者,computer 计算机
	-ician	……家,……能手	technician 技师,mechanician 机械师
	-ist	从事……者	scientist 科学家,chemist 化学家
	-or	……者(人或物)	operator 操作者,censor 传感器
	-acy	性质、状态等	accuracy 精密,determinacy 确定性
	-age	状态、行为等	storage 储存,voltage 电压
	-al	动作、过程等	approval 赞许,removal 移去
	-ance,-ence	性质、状态、行为、过程等	resistance 抵抗,difference 差别
	-ancy,-ency	性质、状态、行为、过程等	constancy 恒定,efficiency 效率
	-bility	动作、性质、状态等	reliability 可靠性,possibility 可能性
	-ety	性质、状态等	variety 变化, dubiety 怀疑
	-faction, -facture	做成,……化,作用等	liquefaction 液化, manufacture 手工制造
	-fication	做成,……化	amplification 放大,simplification 简化
	-ine	表示抽象概念	discipline 学科, machine 机器
	-ing	动作的过程、结果、对象等	reading 读数,building 建筑
	-ion, -sion, -tion, -ation, -ition	行为的过程、结果、状况等	action 作用,conclusion 结论,production 生产,specification 规范,com position 组成
	-ity	性质、状态、程度等	density 密度,reality 现实
	-ment	性质、状态、过程、手段等	movement 运动,treatment 处理
	-ness	性质、状态、程度等	hardness 硬度,slenderness 柔性
	-ship	情况、状态、性质、技巧等	scholarship 常识, relationship 关系

续表

词性	词缀	意义	词例
名词	-th	动作、过程、性质、状态	width 宽度，growth 增长
	-tude	性质、状态、程度	magnitude 量值，latitude 纬度
	-ure	行为、结果	fracture 断裂，pressure 压力
	-graphy	……学、写法等	petrography 岩石学，bibliography 书目
	-ics	……学，……法	dynamics 动力学 bionics 仿生学
	-logy	……学，……论	geology 地质学，hydrology 水文学
	-ant，-ent	产生的物品或物质	resultant 合力，solvent 溶剂
形容词	-able，-ible	可能的，可以的	applicable 能应用的，permissible 容许的
	-al	……的	lateral 横向 additional 附加的
	-ant，-ent	……的	important 重要的，dependent 依赖的
	-ar	……形状的，……特性的	regular 有规则的，linear 线性的
	-ary	属于……的，与……有关的	contrary 相反的，elementary 基本的
	-ive	与……特性的，与……有关的	substantive 本质的，decisive 决定性的
	-ory	属于……的，……性质的	preparatory 预备的，compulsory 强制的
	-ful	充满的，引起……的	plentiful 充足的，useful 有用的
	-ous	充满……的	continuous 连续的，porous 多孔的
	-en	由……制成，……质的	wooden 木制的，earthen 泥土的
	-ble，-ple	……倍的	double 两倍的，quadruple 四倍的
	-fold	倍数	twofold 两倍数的，manifold 多倍的
	-most	最……的	utmost 极度的，topmost 最上的
	-less	没有……的，无……的	wireless 无线的，stainless 不锈的
	-ic，-atic，-ical	属于……的，与……有关的	metallic 金属的，systematic 系统的
动词	-en	使成为，引起	harden 硬化，strengthen 加强
	-fy	致使，使成为	verity 证实，classify 分类
	-ize（ise）	变成，……化	realize 实现，standardize 使……标准化
副词	-ly	状态，程度	relatively 相对地，comparatively 比较地
	-ward（s）	方向	onwards 向前，upwards 向上
	-ways	方向，方式	endways 竖向，sideways 向一边
	-wise	方向，方式	endwise 侧着，lengthwise 顺着

Appendix Ⅱ Expression of Common Mathematical Symbol 常用数学符号的文字表达

1/2	a half, one half
1/3	a third, one third
2/3	two thirds
1/4	a quarter, one quarter, a fourth, one fourth
1/100	a (one) hundredth
1/1000	a (one) thousandth
113/324	one hundred and thirteen over three hundred and twenty four
4(2/3)	four and two thirds
0.25	zero (0, naught) point two five
$+$	plus, positive
$-$	minus, negative
$\pm$	plus or minus
$\times$	multiplied by, times
$\div$	divided by
$=$	be equal to, equal
$\approx$	be approximately equal to, approximately equals
( )	round brackets; parentheses
[ ]	square (angular) brackets
{ }	braces
$\leqslant$	less than or equal to
$\geqslant$	more than or equal to
∞	infinity
$\because$	because
$\therefore$	therefore
$\rightarrow$	maps into
$x + y$	x plus y
$(a + b)$	bracket a plus b bracket closed
$a=b$	a equals b; a is equal to b; a is b
$a\neq b$	a is not equal to b; a is not b
$a\pm b$	a plus or minus b
$a\approx b$	a is approximately equal to b

$a>b$ — a is greater than b

$a\gg b$ — a is much [far] greater than b

$a\geqslant b$ — a is greater than or equal to b

$a<b$ — a is less than b

$a\ll b$ — a is much less than b

$a\leqslant b$ — a is less than or equal to b

$a\perp b$ — a is perpendicular to b

$x\to\infty$ — x approaches infinity

$a\equiv b$ — a is identically equal to b; a is of identity to b

$\angle a$ — angle a

$a\parallel b$ — a is parallel to b

$a\sim b$ — a varies directly as b

$a\backsim b$ — the difference between a and b

x^2 — x square; x squared; the square of x; the second power of x; x to second power

x^3 — x cube; x cubed; the cube of x; the third power of x; x to the third power

$\sqrt{x}$ — the square root of x

$\sqrt[3]{x}$ — the cube root of x

% — per cent

2% — two per cent

‰ — per mill

5‰ — five per mill

$\log_n{}^x$ — log x to the base n

$\log_{10}{}^x$, $\lg x$ — log$x$ to the base 10, common logarithm

$\log_e{}^x$, $\ln x$ — log x to the base e, natural logarithm, n apiarian logarithm

e^x — exponential function of x, e to the power x

x^n — the nth power of x, x to the power n

$x^{1/n}$ — the nth root of x, x to the power one over n

sin — sine

cos — cosine

tg, tan — tangent

ctg, cot — cotangent

sc, sec — secant

csc, cosec — cosecant

$\sin^{-1}$, arcsin — arc sine

$\cos^{-1}$, arccos — arc cosine

sin$h$ — the hyperbolic sine

cos$h$ — the hyperbolic cosine

Σ — the summation of

$\sum_{i=1}^{n} x_i$	the summation of x sub i, where i goes from 1 to n
Π	the product of
$\prod_{i=1}^{n} x_i$	the product of x sub i, where i goes from 1 to n
$\mid X \mid$	the absolute value of x
$\overline{x}$	the mean value of x; x bar
b'	b prime
b''	b double prime; b second prime
b'''	b triple prime
$f(x)$	function f of x
Δ	finite difference or increment
Δx, δx	the increment of x
dx	dee x; dee of x; differential x
dy/dx	the differential coefficient of y with respect to x; the first derivative of y with respect to x
$d^2 y/dx^2$	the second derivative of y with respect to x
$d^n y/dx^n$	the nth derivative of y with respect to x
$\partial y/\partial x$	the partial derivative of y with respect to u, where y is a function of u and another variable (or variables)
$\int$	integral of
$\iint$	double integral of
$\int \cdots \int$	n-fold integral of
$\int_a^b$	integral between limits a and b (... from a to b)
F	vector F
a^2	a sub two
20°	twenty degrees
7′	seven minutes; seven feet
13″	thirteen seconds; thirteen inches
0℃	zero degree Centigrade(Celsius)
100℃	one (a) hundred degrees Centigrade
32℉	thirty-two degrees Fahrenheit

Appendix Ⅲ Matrixing of Unit and Length, Capacity and Weight in Civil Engineering

土木工程中常用的度量衡和单位换算

	The Metric System 公制		GB & US System 英美制	
	英文、中文名称	简写	英文、中文名称及换算	简写及换算
长度 Length	1 centimeter 厘米	cm	0.3937 inches 英寸	1 in=2.54cm
	1 meter 米	m	3.2808 feet 英尺 1.0936 yards 码	1 ft=0.3048m 1 yd=0.9144m
	1 kilometer 千米	km	0.6214 miles 英里	1 mi=1.6093km
面积 Area	1 square millimeter 平方毫米	mm^2	0.00155 square inches 平方英寸	1 sq. in. = 645.16mm^2
	1 square meter 平方米	m^2	10.7643 square feet 平方英尺 1.196 square yards 平方码	1 sq. ft. =0.0929m^2 1 sq. yd. =0.836m^2
	1 square kilometer 平方千米	km^2	0.3861 square miles 平方英里	1 sq. mi=2.59km^2
体积 Volume	1 cube meter 立方米	m^3	35.3357 cube feet 立方英尺 1.308 cube yards 立方码	1 cu. ft. =0.0283m^3 1 cu. yd. =0.7645m^3
重量，力 Weight, fo-rce	1 kilogram 千克	kg (kgf)	2.2046 pounds 磅	1lb(lbf)=0.4536kg
	1 tone 公吨	t (tf)	0.9842 long tons 英吨 1.1025 short tons 美吨 9.8067 kN(千牛)	1 long ton=1.106 tf= 9.964kN 1 short ton=0.907 tf= 8.896tf
	1 Newton 牛顿(国际单位)	N	0.2248 pounds 磅	1 1b(1bf)=0.4482N 1 N=0.102kgf 1 kgf=9.8066N
	1 kilo Newton 千牛	kN	0.2248 kips 千磅	1 kip=4.4482kN
速度 Velocity	1 meter/second 米/秒	m/s	2.2369 miles/hour 英里/小时	1 mi/hr=0.447kN
	1 kilometer/hour 千米/小时	km/h	0.6214 miles/hour 英里/小时	1mi/h=1.6093km/h
压强，应力 Pressure Stress	1 per square meter 牛顿/平方米	N/m^2	0.000145 pounds per square inch 磅/英寸^2	1 psi=6894.76N/m^2
	1 million Newton per square millimeter 兆帕	MPa	0.145 kilopounds per square inch 千磅/英寸^2	1ksi=6.895MPa 1MPa =10.204kgf/cm^2
力的线集度 Linear Load	1 kip/ft=1488.16kgf/m=14.59kN/m 1 ton/m=0.672kip/ft， 1kN/m=0.0685kip/ft			

References

[1] 汪德华. 建筑工程专业英语. 北京：地震出版社，2003.

[2] 郭向荣，陈政清. 土木工程专业英语. 北京：中国铁道出版社，2001.

[3] 李嘉. 专业英语（土木工程专业道桥方向）. 北京：人民交通出版社，2003.

[4] 李亚东. 新编土木工程专业英语. 成都：西南交通大学出版社，2000.

[5] 段兵廷. 土木工程专业英语. 湖北：武汉工业大学出版社，2000.

[6] 赵明瑜. 土木建筑系列英语. 北京：中国建筑工业出版社，1987.

[7] 王建成. 科技英语写作. 西安：西北工业大学出版社，2000.

[8] 邓贤贵. 建筑工程英语（第二版）. 武汉：华中理工大学出版社，1997.

[9] 建筑抗震设计规范 GB 50011—2010. 北京：中国建筑工业出版社，2016.

[10] 混凝土结构设计规范 GB 50010—2010. 北京：中国建筑工业出版社，2015.

[11] 建筑结构设计术语和符号标准 GB 50083—2014. 北京：中国建筑工业出版社，2014.

[12] 建筑结构荷载规范 GB 50009—2012. 北京：中国建筑工业出版社，2012.

[13] 砌体结构规范 GB 50003—2011. 北京：中国建筑工业出版社，2011.

[14] 建筑地基基础设计规范 GB 50007—2011. 北京：中国建筑工业出版社，2011.

[15] 高层建筑混凝土结构设计规程 JGJ 3—2010. 北京：中国建筑工业出版社，2010.

[16] Chu-Kia Wang & Charles G. Salmon. Reinforced Concrete Design，New York：Harper International Edition. Harper & ROM Publishers，1979.

[17] Augustine I. Fredrich. Sons of Martha：Civil Engineering Readings in Modern Literature. New York：ASCE，1989.

[18] Raymond Sterling. Underground Space Design. Von Nostrand Reinhold，1993.

[19] Leo Diamant and C. R. Tumblin. Construction Cost Estimates. Second Edition. John Wiley & Sons I nc.,1990.

[20] James Wines，Philip Jodidio. Green architecture. Köln. New York：Taschen，2000.

[21] Brian Edwards. Green architecture. Chichester：Wiley-Academy，2001.

[22] Ryan E. Smith. Prefab architecture：a guide to modular design and construction. Hoboken，N. J.：John Wiley & Sons，2010.

[23] Sheri Koones. Prefabulous sustainable building and customizing an affordable，energy-efficient home. New York：Abrams，2010.

[24] Barry James. Sullivan. Industrialization in the building industry. New York：Van Nostrand Reinhold，1980.

[25] James E. Ambrose，Patrick Tripeny，Harry Parker. Simplified design of wood structures. Hoboken，N. J.：Wiley，2009.

[26] Marc Wilhelm Lennartz，Susanne Jacob-Freitag. New architecture in wood：forms and structures. Basel：Birkhäuser，2015.

[27] D. A Nethercot. Composite structures，London. New York：Spon Press，2003.

[28] European Convention for Constructional Steelwork. Technical General Secretariat. Composite structures，London. New York：Construction Press，1981.

[29] R. P. Johnson（Roger Paul）. Composite structures of steel and concrete. London：Granada，1975.